The Petroleum Papers

Geoff Dembicki

THE Petroleum Papers

Inside the Far-Right Conspiracy to Cover Up Climate Change

DAVID SUZUKI INSTITUTE

GREYSTONE BOOKS

Vancouver/Berkeley/London

22 23 24 25 26 5 4 3 2 1

Greystone Books Ltd.
greystonebooks.com

David Suzuki Institute
davidsuzukiinstitute.org

Cataloguing data available from Library and Archives Canada
ISBN 978-1-77164-891-2 (cloth)
ISBN 978-1-77164-892-9 (epub)

Editing by David Beers
Copy editing by Paula Ayer
Proofreading by Alison Strobel
Indexing by Stephen Ullstrom
Jacket and text design by Jessica Sullivan
Jacket photograph by Creative Images / Alamy Stock Photo

Printed and bound in Canada on FSC® certified paper at Friesens. The FSC® label means that materials used for the product have been responsibly sourced.

Greystone Books thanks the Canada Council for the Arts, the British Columbia Arts Council, the Province of British Columbia through the Book Publishing Tax Credit, and the Government of Canada for supporting our publishing activities.

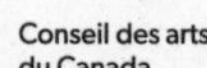

Canada Council for the Arts Conseil des arts du Canada

Greystone Books gratefully acknowledges the xʷməθkʷəy̓əm (Musqueam), Sḵwx̱wú7mesh (Squamish), and səl̓ílwətaʔɬ (Tsleil-Waututh) peoples on whose land our Vancouver head office is located.

For Kara and Yoko

Contents

V: BLAME CANADA (2006–2010)

VI: THE CLIMATE GOES TO COURT (2008–2014)

VII: WELL-OILED ALLIES (2016–2019)

VIII: THE RIGHT TO LIVE (2020–2022)

Introduction

MIDWAY THROUGH a historic congressional hearing about whether some of the world's largest oil companies lied to the public about climate change, I noticed something odd. It was October 28, 2021, and I was watching the event through a livestream on my laptop in my small Brooklyn apartment. The House Committee on Oversight and Reform had that day hauled in the heads of Exxon, BP, Shell, Chevron, and the American Petroleum Institute for questioning. Democrats wanted answers about a multidecade campaign led by the oil and gas industry to convince millions of people that the climate emergency doesn't exist—and that even if it does, there is no point in fighting it. But Republicans had called in their own witness to testify: a welding foreman from Fouke, Arkansas. Neil Crabtree was there to say he'd lost his job on a canceled pipeline that would have stretched from the Canadian province of Alberta to the Texas Gulf Coast. What did that have to do with a hearing in Washington, D.C., about Big Oil disinformation?

Like the dozens of other climate journalists covering this event, I felt pressure to get the facts right. At the same moment that oil executives were being grilled, the Joe Biden administration was trying to pass a multitrillion-dollar spending package through Congress that contained tax credits supporting clean energy and electric vehicles, penalties for harmful methane emissions, and a policy pushing electric utilities to phase out natural gas and coal. If this "Build Back

Better" Act became law, it would be the most substantial action ever taken by the United States on climate change.

I knew from reporting about climate change for more than a dozen years that during moments of great possibility, the roar of disinformation is at its loudest. This book will chart many such high-stakes moments. And now another one had certainly arrived. "This is our last best chance to pass meaningful climate legislation," Danielle Deiseroth, senior climate analyst at the progressive polling organization Data for Progress, told me of the Biden agenda. "If we delay action any longer, we're going to be really screwed."

Finally, at least, oil executives were being called out for their tactics by America's top lawmakers. "Today, the CEOs of the largest oil companies in the world have a choice: you can either come clean... and stop supporting climate disinformation, or you can sit there in front of the American public and lie under oath," said Democratic representative Ro Khanna. The CEOs answered impassively. "I don't believe companies should lie, and I would tell you that we don't do that," Exxon's Darren Woods said in response to Khanna.

Amid the rich, polished executives, Crabtree couldn't help but stand out. The pipeline worker from Arkansas wore his suit and tie awkwardly. His voice wavered as he explained that he had been hired to help build the Keystone XL pipeline, a project designed to bring 830,000 barrels of oil per day from Canada into the United States. Because this oil would have been derived from an especially polluting petroleum deposit called the Canadian oil sands, which are sometimes referred to as "tar sands," Keystone had been the target of years of environmental protests. Biden canceled the project on his first day in office.

"I lost my job," Crabtree said. "There seems to be no thought given to the hundreds of thousands of workers in this industry or the millions of products that we use every single day that are provided by fossil fuels."

Crabtree presented a sympathetic figure, an honest and hard-working American seemingly chewed up by the political machinations of Washington. But I sensed there was more to his story.

With the hearing on in the background, I searched Crabtree's name. One of the first results was the website for a group called Americans for Tax Reform. On the site there was a self-recorded video of Crabtree inside a vehicle. Instead of a suit, he was dressed in a hoodie and baseball cap. "Canceling this Keystone Pipeline to make a group of people happy has had real life consequences," he said. "We got people who can't work now, can't provide for their families." I saw that underneath the video there was a special request that people send more such videos: "Americans for Tax Reform is collecting personal testimonials of Americans hit by President Biden's policies." Crabtree's appearance at the hearing was starting to make more sense.

That's because Americans for Tax Reform isn't just some ordinary conservative group. It was one of the authors of the climate change denial playbook. The organization helped create an "action plan" in 1998 along with Exxon, BP, Shell, Chevron, the American Petroleum Institute, and others to flood mainstream media with disinformation about the scientific consensus on global warming. A leaked document lays out the strategy in detail: "Develop and implement a national media relations program to inform the media about uncertainties in climate science; to generate national, regional and local media coverage on the scientific uncertainties, and thereby educate and inform the public, stimulating them to raise questions with policymakers."

Not much had changed in the nearly twenty-five years since then. Not long after Joe Biden was elected on a platform to invest trillions in green industries, create millions of new jobs, and drastically lower U.S. emissions, the head of Americans for Tax Reform, a well-known Republican operative named Grover Norquist, coauthored an op-ed in the *Washington Examiner* declaring war on "the extreme

Left's big-government climate agenda." Crabtree's testimonial was part of a carefully coordinated media campaign that Norquist dubbed #BidenKilledMyJob.

Several months later, Crabtree was broadcasting live as a witness at the House congressional hearing on climate disinformation. After he was done, as CEOs like Exxon's Woods were asked to account for spreading scientific denial, Republicans steered the conversation back to Crabtree. "It's critical that this committee examine the pressing concerns of American citizens," said James Comer, a Republican Congress member from Kentucky who once said, "I do not believe in global warming." Crabtree's remarks were carried by mainstream outlets like the *New York Times*, CNN, the *Guardian*, Reuters, the BBC, and the *Daily Mail*, but none highlighted his ties to the machinery of Big Oil's climate crisis denial.

There is no reason to suspect that Crabtree was lying about losing work on Keystone XL—and he might not have even been aware of Americans for Tax Reform's history of denying climate change. But the media attention his testimony received was a victory for Republicans, the oil and gas CEOs testifying under oath, and the Canadian oil sands industry. Once again, these forces were intentionally distracting the public from the bigger picture: that the stable climate upon which all human life depends is being altered beyond recognition.

It was disheartening to see this playbook still being used decades after its creation, because my research had educated me to how, time and again, such tactics laid waste to humanity's easiest, best chances to fend off climate catastrophe. Now I could see nothing but lost years, a window of opportunity spanning more than half a century, repeatedly and forcefully shut by powerful interests who've known and suppressed the truth.

Many people take as a given the perilous point at which our planet has arrived. They struggle to imagine a world in which fossil fuels aren't driving us toward the precipice. But my reporting has made me realize there was nothing inevitable about the chaotic future we face, with all of the natural systems that support us collapsing. It did not have to come to this. We are here because at each critical juncture, fossil fuel companies, dependent on a particularly dirty form of oil, worked in concert with political allies in the United States and Canada to build an edifice of lies that has prevented our self-rescue.

I know this—and will show it in this book—because virtually all the evidence is publicly available. But the details have been scattered haphazardly across the internet, or else contained in vast digital archives—hosted by watchdog groups such as the Climate Investigations Center, DeSmog, the Center for International Environmental Law, Koch Docs, and others—that most people don't have the time or know-how to peruse. Several years ago, I decided to dive into this ocean of information headfirst. Since then, I've navigated hundreds of documents, firsthand accounts, news pieces, and studies. I've done in-depth interviews with key players: those that helped build the disinformation machine, those who've made it their life's work to understand it, and those now trying to tear it down. And through it all, I have had to fight my own rising waves of anger at what I was discovering and piecing together.

These petroleum papers weave together a wide cast of fascinating characters. They tell of Robert Dunlop, a U.S. oil executive who disregarded a dire warning about climate change in 1959 from the inventor of the atomic bomb and instead tapped a massive new source of oil in Canada. They reveal how Exxon, Koch Industries, Shell, and others rushed to exploit the oil sands, even as they privately learned it would crank up the climate's heat and cause "catastrophic" death and destruction. They show these companies quietly anticipated public outcry about climate change, and prepared a strategy to neutralize it, so that when NASA scientist James Hansen brought the emergency to the public's attention in 1988, the oil industry was able to quickly sow doubt via media campaigns.

Contained in the papers are figures everyone will know—Rupert Murdoch, Donald Trump—and some names immediately recognized only on one side the border—Canada's former prime minister

Stephen Harper, for example, a one-time employee of an oil company owned by Exxon. This saga of power and wealth pursued at the expense of a survivable planet includes beltway lobbyists, astroturf organizers, and politicians for whom oil was literally a religion.

This story includes, as well, people fighting back. A famous class-action lawyer who helped convict the tobacco industry for lying about cancer and is now using the courts to reveal Big Oil's lies. And a formerly trusted Exxon employee who was pushed out of the company for asking too many hard questions about the company's climate change denial. My journey took me from Seattle to Washington, D.C., to the oil sands themselves. It also took me to the Philippines, where I met a young woman who knows, all too excruciatingly well, where our world has arrived after so many destroyed opportunities.

I'll start the story with her.

I

The First Warnings

(1959–2013)

“Just another storm”

JOANNA SUSTENTO woke up that morning with a feeling of pleasant expectation. She was a Filipina in her early twenties getting started in a career in business and living with her family in Tacloban City, a coastal city on the central Philippines island of Leyte. Sustento shared the small space with her dad, a lawyer; her mom, the manager of a government bank; her two older brothers; her sister-in-law; and her nephew, a three-year-old named Tarin.

Apart from Tarin, Sustento was the youngest person there, “the only girl in the family,” she said. “So being the only girl, my parents, especially my dad, would often call me ‘princess.’ I didn’t really like it when he called me that.”

The house was crowded yet infused with warmth. In a culture where family is considered the foundation of all social life, Sustento held many happy memories of large get-togethers of extended relatives. That past week, in fact, had been a blur of communion. After many years away, several relatives on her father’s side had decided to visit Tacloban, resulting in a packed schedule of “family gatherings and just hanging out and going to family lunches and dinners.”

No one had been paying too much attention to the warnings on the news that a tropical storm, known in the South Pacific as a typhoon, was headed toward the city. This on its own wasn’t necessarily alarming news for Tacloban’s inhabitants. “We experience an average of twenty typhoons a year,” Sustento said. “They’re just normal for us.”

Tacloban's 240,000 or so residents occupy a narrow stretch of land bordered to the east by San Pablo Bay and to the west by Mount Naga-Naga. According to its official slogan, Tacloban is "a city of progress, beauty and love." It's also a city of low incomes, even by Filipino standards. Over 40 percent of people in the region of Eastern Visayas, where Tacloban is located, lived below the poverty line in 2012. Many of them resided in small homes built out of concrete and corrugated steel located on or close to the coastline. The vast majority of homes and buildings were less than sixteen feet above sea level, and with no coastal defenses, they were highly exposed to storms and floods.

The news reports that week stressed that this approaching storm was much stronger and potentially more dangerous than usual. But hearing about above-average wind and rain intensity felt like an abstract and distant threat: "You couldn't really wrap your head around it," Sustento said.

The night before the storm was supposed to hit, her family was over for dinner at her grandmother's house. "When we were about to go home I had this uncle from a different city talking to my dad suggesting for us not to go home and to wait the storm out until it's over," she said. "But of course my dad insisted on going home." Having experienced so many typhoons in the past, they figured "our neighborhood would be safe, our house would withstand the storm."

Sustento was even a bit excited. Since she was a little girl, she'd loved typhoons. A tropical storm hitting Tacloban meant that the city's busy street life—its downtown crowds shopping for iPhones and vegetables, people eating cheeseburgers at Jollibee's, roadside vendors barbecuing pork skewers, drivers of brightly colored Jeepneys whisking people north and south—came to a halt. "It would mean that classes would be suspended for days, there would be no work," she said. "So I would just be at home with my family making the most of the cozy, chilly weather." Occasionally a storm would knock out power lines. "There would be blackouts, we'd be out of power for days," she said. But that just added to the experience.

Now, early in the morning of November 8, 2013, Sustento was roused out of her sleepy reverie by rain coming in through her open bedroom window. She got up and tried to close it, "but I had a hard time... because the wind was so strong." Her father was outside and Sustento went out to join him, looking toward the horizon. "Everything was so white," she said. "It was very hazy." Back indoors, the family prepared breakfast. "My dad told me to make some coffee." Meanwhile three-year-old Tarin ran around the house. "I could see the joy in his eyes, probably because we were all together."

But as the wind picked up Sustento started to feel a little bit nervous. The whole house felt like it was vibrating. "Then I hear this very eerie sound. The wind was howling... It was as if a monster were coming out from the skies." Sustento opened a screen door to peer outside and the "intensity was so strong my lips were being blown away as I was speaking."

She ducked back inside, looked up, and noticed water dripping from the ceiling. Sustento went to get a basin, but along the way she saw that water was coming in through the kitchen door. "Thinking it was rainwater, I covered the doors with the rugs," she said. "Then I saw that water was also coming inside from our main door, so again we tried to cover that with the rugs, thinking it would absorb it. But it was already starting to flood." Looking at the clock, she saw that it was nearly 7:00 AM. Sustento kept glancing at her parents. She was "looking for clues" from their facial expressions, something to tell her if she should "be scared or if everything is still okay or normal." By that point the typical laughter and morning chatter had stopped. "It was unusual for them to stay silent."

The typhoon had begun five days earlier as a low-pressure system hundreds of miles southeast of Tacloban. As it gathered force, the Japan Meteorological Agency upgraded it to a tropical depression, and weather authorities issued a storm warning for the South Pacific island nations in its path. Not long after that, the storm received official typhoon designation and a name: Haiyan. On November 6, Typhoon Haiyan slammed into Palau and Micronesia with 150-mile-per-hour winds, destroying homes and knocking out power and water supplies. Helicopter pilots saw trees and buildings scattered like twigs across the landscape.

Picking up energy and intensity, Haiyan headed toward the Philippines, where local authorities assigned it the name Yolanda. It's common in this part of the world for typhoons to receive both an international name and a local one. By this point Haiyan had become the most powerful storm recorded the entire year. "Let us do all we can while [the typhoon] has not yet hit land," then Philippines president Benigno Aquino said in a speech broadcast on national TV the day before the advancing threat made landfall. "Let us remain calm, especially in buying our primary needs, and in moving to safer places." National experts were getting nervous, however. "This is a very dangerous typhoon," state weather forecaster Glaiza Escullar said. "There are not too many mountains on its path to deflect the force of impact, making it more dangerous."

Some of the danger could also be attributed to climate change. That year, 2013, was the fourth-warmest year since records started in 1880, with the planet's combined land and ocean surface temperature 0.62 degrees Celsius higher than the average of 13.9 degrees (57°F) recorded during the twentieth century. This was due largely to global emissions of carbon dioxide—many of them caused by the burning of coal, oil, and gas. Those emissions rose to 35.8 billion tons in 2013, a 60 percent increase when compared to the year 2000.

The Philippines has always been highly exposed to typhoons because it's surrounded by water warmed by the equatorial sun. This water releases some of the heat into the atmosphere, which then creates wind and rain clouds. "However, as the ocean's surface temperature increases over time from the effects of climate change, more and more heat is released into the atmosphere," the Climate Reality Project, a nonprofit organization founded by Al Gore, explained. "This additional heat in the ocean and air can lead to stronger and more frequent storms—which is exactly what we've seen in the Philippines over the last decade."

On the evening before Haiyan hit Sustento's home, the typhoon was comparable in area to the entire land mass of the Philippines. Depicted in real time on meteorological maps, it resembled a massive buzz saw slicing toward Tacloban—and with sustained winds of almost two hundred miles per hour, it was equivalent to a Category 5 hurricane, the same designation given to Hurricane Andrew, which in 1992 caused sixty-five deaths and $27.3 billion in damage when it tore through the Bahamas, Florida, and Louisiana.

CNN reported live on the approaching typhoon. "This is probably one of the top dozen of all storms ever seen on this planet," the network's chief meteorologist, Chad Myers, said on air. "Chad," said CNN anchor Suzanne Malveaux during the broadcast, "is there anything—in light of the fact that we've got this warning, we know it's coming—is there anything people can do there, because I imagine the damage could be extraordinary if people don't pay attention or at least don't prepare for what's going to hit them."

"We will lose people no matter what you can do," Myers replied. "This is almost at some point for some people on these islands an unsurvivable storm."

But the deadly chaos barreling toward Sustento and her family wasn't just the result of chance and bad luck. Unbelievable as it might seem, the events that led to the most powerful storm in history were set in motion more than fifty years earlier.

"Men on a hunt"

ON A NOVEMBER DAY in 1959, one of the most powerful people in oil and gas received a credible warning that his industry could cause death and suffering for large numbers of the planet's inhabitants. The warning came during an event in New York City called "Energy and Man," a daylong symposium that was attended by government officials, historians, economists, scientists, and fossil fuel industry executives—more than three hundred in total. Organized by the American Petroleum Institute along with Columbia University's Graduate School of Business, the event was described as "an inquiry into the role of energy, past, present, and future, in the lives of each of us." It was also a celebration of the industry's one hundredth year of existence.

Robert Dunlop was among the attendees that day who walked up the stone steps of Columbia's Low Library, passed by Ionian columns and bronze busts of Zeus and Apollo, entered a foyer containing a white marble statue of Athena, and entered a library reading room lined with green marble columns topped with gold. Dunlop presented himself as the successful fifty-year-old businessperson that he was: black tie, horn-rimmed glasses, not a trace of stubble, and hair slicked back and meticulously parted to the side.

He was the second speaker at this gathering of postwar power brokers, and he opened his thirty-five-minute speech with a not-so-subtle metaphor. "It pleases me to note that on its hundredth birthday, the petroleum industry is not a tired, old industry. The petroleum revolution is not history, but is a current, pulsating event, still unfolding," he told the crowd lasciviously. "Under the thrust of competition, which was midwife at its birth, petroleum developed as an industry of great independence and vitality, as a restless innovator which continuously shook up the status quo."

Born in Boston, Dunlop attended the Wharton School of Finance and Commerce at the University of Pennsylvania and graduated at the top of his class in 1931. He was soon after hired as an accountant by Sun Oil, at the time a large regional oil company in the U.S. northeast. Sixteen years later, at the age of thirty-seven, he was named president, and under his leadership Sun Oil expanded aggressively across the United States and into Canada and Latin America. "Endowed with remarkable recall—always remembering the names of employees after an initial meeting—Dunlop was serene and pensive, as well as sometimes self-effacing, often referring to his accounting roots but obviously proud of his superb technical prowess," his alma mater would later describe him.

Dunlop was at Columbia that day not only as head of Sun Oil but as a director of the American Petroleum Institute, a trade association founded in 1919 which by the late 1950s had become the main research and lobbying group for the U.S. oil and gas industry. It was also a leading source of fossil fuel propaganda. A black-and-white film produced by the institute during this period showed two oil company geologists stepping off a helicopter onto an Oklahoma prairie potentially containing new supplies of petroleum. "You are looking at men on a hunt," the narrator explains. "The game they seek is not a living thing, and their weapons are not firearms, but they are hunting, and they are armed—with the weapons of science."

During his speech that day, Dunlop sought to describe the challenges of early oilmen like Edwin Drake, who had drilled the first successful U.S. oil well in 1859. His phallic metaphors now turned biblical. "At least one man was purported to have argued with Colonel Drake that his idea was immoral; that the oil was needed down there for the fires of hell, and to withdraw it was to protect the wicked from the punishment which they so just deserved," Dunlop said. "While this proved not to be the last criticism directed against the petroleum industry, it was at least among the most novel."

A century after the industry's founding, Dunlop projected confidence about the decades ahead. "To me, the future appears to hold great opportunities for the oilman," he said in New York. However, Dunlop hinted cryptically at challenges to come: "There are aspects of the future which are clouded by the penetration of noneconomic forces into the functioning of an industry which has always performed best in an atmosphere of economic freedom." Dunlop didn't say what storm clouds specifically threatened the expansion of oil and gas. But the next Energy and Man speaker did. That speaker vividly described a new and unexpected threat to the industry: its vast and growing emissions of a greenhouse gas called carbon dioxide.

Edward Teller was no back-to-nature romantic. The scientist who took the podium after Dunlop was known primarily for his lead role in creating the world's first weapons of mass destruction. Teller was part of a secret government team at Los Alamos, New Mexico, that during the 1940s developed atomic bombs that were dropped on Hiroshima and Nagasaki, causing the deaths of over 200,000 people. Yet Teller began his talk at Columbia by warning of a threat that was potentially greater than nuclear destruction. "And this, strangely, is the question of contaminating the atmosphere," he said.

Teller tried to ease his audience into a concept many of them were likely hearing for the first time. "Now," he explained, "all of us are familiar with smoke and smog and all of us know about it as a nuisance." Teller went on, "I would like to talk to you about a more hypothetical difficulty which I think is quite probably going to turn out to be real. Whenever you burn conventional fuel, you create carbon dioxide. It has been calculated that the carbon dioxide which has been put into the atmosphere since the beginning of the industrial revolution equals approximately 10 percent of the amount of carbon dioxide that our atmosphere contained originally."

This was a fairly avant-garde statement for the time, but not unprecedented. Scientists had first described the ability of gases in the atmosphere to trap planetary heat during the 1800s, and in the early 1900s Swedish chemist Svante Arrhenius was one of the first scientists to link this warming to human activity, calculating the greenhouse gases released by burning coal. By the 1950s, the science was becoming more precise through the work of people like Roger Revelle at the Scripps Institution of Oceanography, who along with Hans Suess of the U.S. Geological Survey showed the oceans are limited in their ability to absorb carbon dioxide. Most of the gases released through our use of fossil fuels, they calculated, end up in the atmosphere.

Depictions of global warming's impact on human society were also starting to enter popular culture. Several years before Teller attempted to alert his New York audience, an article in *Time* magazine warned that CO_2 building up in the atmosphere could by the early 2000s "have a violent effect on the earth's climate."

It's no coincidence that Teller was aware of the latest climate science. During the same period when it was funding early atomic research, the U.S. military was also looking closely into earth sciences. This included Project Gabriel, a classified study of the impact of nuclear weapons on weather patterns launched by the U.S. Weather Bureau in 1949. In subsequent studies, military scientists sought to figure out through models the best weather conditions for explosions and how detonating nuclear weapons could affect the natural world. It was during the course of such research that these scientists first coined the term "environmental sciences."

Teller knew the audience at Columbia might find it hard to be too concerned about carbon dioxide, a greenhouse gas that can't be seen by the human eye and poses no immediate danger to people's health. "So why should one worry about it?" he asked. Teller then provided an intro to the burgeoning field of climate science: "[CO_2's] presence in the atmosphere causes a greenhouse effect in that it will allow the solar rays to enter, but it will to some extent impede the radiation from the Earth into outer space. The result is that the Earth will continue to heat up until a balance is re-established."

The impacts, he warned, could be catastrophic. "It has been calculated that a temperature rise corresponding to a 10 per cent increase in carbon dioxide will be sufficient to melt the icecap and submerge New York," he said. "All the coastal cities would be covered, and since a considerable percentage of the human race lives in coastal regions, I think that this chemical contamination is more serious than most people tend to believe."

This wasn't the first time Teller had delivered such a warning. In late 1957, he'd spoken to members of the American Chemical Society about the impact of burning coal and oil, stating that the carbon emissions coming out of chimney and exhaust pipes could someday change the climate and melt a large portion of the polar ice caps. This had apparently caught the oil industry's attention. In October 1959, the month before Teller's Columbia speech, Royal Dutch Shell scientist Dr. M. A. Matthews published a paper saying he didn't think Teller's prediction would come true because "nature's carbon cycles are so vast that there seem few grounds for believing Man will upset the balance."

Yet there is reason to think many of the oil executives and industry representatives hearing Teller's speech in New York were learning about climate change for the first time. "Could you please summarize briefly the danger from increased carbon dioxide content in the atmosphere?" the dean of the Columbia Graduate School of Business, Courtney Brown, asked following Teller's talk.

"Our planet will get a little warmer," the nuclear scientist responded, explaining that global CO_2 levels would grow exponentially if we continued to burn petroleum and other fossil fuels. "It is hard to say whether it will be two degrees Fahrenheit or only one or five. But when the temperature does rise by a few degrees over the whole globe, there is a possibility that the icecaps will start melting and the level of the oceans will begin to rise."

"Well," Teller went on, "I don't know whether they will cover the Empire State Building or not, but anyone can calculate it by looking at the map and noting that the icecaps over Greenland and over Antarctica are perhaps five thousand feet thick."

There is no historical record of how Dunlop or any other oil industry executive in the audience reacted. But the Sun Oil CEO was apparently paying close attention, having promised during his own speech to "listen with the keenest attention this afternoon to Dr. Teller." What Dunlop heard that day is one of the earliest known climate change warnings to the oil and gas industry. The products his industry hunted, extracted, and sold, the oil executive now knew, were capable of someday flooding the very city where his industry's hundredth birthday celebration was taking place.

"A gift from God"

IT WAS A WET AND MUDDY September day when Robert Dunlop crossed his nation's northern border to visit what would decades later become one of the biggest oil and gas operations in the world. In 1963, however, the site was a remote patch of boreal forest in northeastern Alberta. It had been nearly four years since the Columbia University Energy and Man conference, and if Dunlop had taken Teller's warning about climate change seriously, his actions didn't reflect it. During those intervening years one of Dunlop's top priorities as president of Philadelphia-based Sun Oil was to figure out a way to commercially develop the Canadian oil sands, a deposit that by some estimates contained more petroleum than all of Texas, and even the entire Middle East.

It had rained the night before Dunlop arrived at Sun Oil's site, making the muskeg difficult to traverse. The only way to navigate this boggy landscape was in a tank-like vehicle known as a

Bombardier. A company record from the visit explains that "boots and galoshes were the order of the day." But not, apparently, for Dunlop. The executive took a walk around the future Great Canadian Oil Sands site in leather dress shoes, wearing a suit and tie with a long overcoat. Nearby, a Sun Oil photo shows, the Bombardier was caked in thick streaks of mud.

It was not yet possible to produce much oil at the site, but even getting to this stage had been a slog. Sun Oil's lease was located more than three hundred miles north of Edmonton, itself the most northerly major city in North America. The company, explains University of Alberta PhD researcher Hereward Longley, "had to build mostly from scratch all of the roads, bridges, buildings, generators, and housing necessary to access and develop the project." Earlier that spring, Sun Oil had brought a tractor north from Mildred Lake, and for two months struggled to cut passageways for heavy machinery through the swamp surrounding the lease. Each new building or drilling platform required cutting down trees and draining wetlands. The six-month-long winter, when temperatures could drop below minus fifty degrees, brought a whole different set of challenges.

People had known about the presence of oil in this region for centuries—if not millennia. In 1715, an official at the Hudson's Bay Company heard a report from a Chipewyan First Nations woman named Thanadelthur likely referring to the Athabasca River, from whose banks oozed a substance known as bitumen, a thick substance of oil mixed with tarry sand. "They tell us there is a certain gum or pitch that runs down the river," the official recorded in his notebook. Later in the century, the fur traders Peter Pond and Alexander Mackenzie described "bituminous fountains" where the Clearwater River meets the Athabasca in northeastern Alberta.

Additional descriptions of the region's vast oil reserves appeared sporadically. But it wasn't until 1914 that a British geologist named Dr. T. O. Bosworth traveled to the region to officially survey the so-called Tar Sands District for potential oil extraction. He deemed it "an admirable oil generating formation." Actually taking the oil out of the ground wouldn't be easy, however. "The oil embedded in the bitumen around Fort McMurray had to be laboriously separated and even then its sulphur-laden content required more refining than the standard product pulled directly from underground sources," explains historian Graham Taylor in *Imperial Standard: Imperial Oil, Exxon, and the Canadian Oil Industry From 1880*.

Because of these difficulties, oil companies mostly stayed away, and Canadian policymakers figured the gigantic oil reserves, estimated around 170 billion recoverable barrels, would mostly be of use as a raw material for road construction. After surveying the area on behalf of the federal Department of Mines, an engineer named Dr. Sidney Ellis decided to cash in on that vision, forming an asphalt company in the 1920s with American businessman Thomas Draper. But the allure of vast oil wealth didn't go away, and in 1924 Alberta scientist Dr. Karl Clark invented a process for separating the oil from bitumen solids that in theory could make the development of a large commercial oil sands project possible.

The day Dunlop got his shoes sticky with mud walking through northern Alberta's muskeg, he was following in the footsteps of his predecessor as Sun Oil president, J. Howard Pew. Pew was not known for his warmth or charisma. One U.S. senator referred to him as a "stiff-necked, bushy-browed, six-footer" with the personality of "an affidavit." The oilman was deeply religious and fervently libertarian, qualities he shared with his brother, Joseph N. Pew Jr. Pew spent much of the 1930s warring against president Franklin D. Roosevelt's New Deal. That economic stimulus effort may have lifted the United States out of the Depression, but Pew saw it as a tyrannical attempt to impose communism.

"The persistent effort to bring industry, business, commerce and enterprise under government domination is a flat denial of all the lessons of the century and a half of the industrial age," Pew argued. The parallels between Pew's reactionary political views and those of a later duo of anti-government billionaires, the Koch brothers, isn't coincidental, according to University of Notre Dame professor Darren Dochuk. "The Kochs," he writes, "are just the latest in a long line of oil-rich brothers driving the Republican Party's rightward march. The very first were the Pews, who, between the 1930s and 1960s, spent their oil fortune remaking the GOP in their libertarian and conservative Christian image."

During that era, Pew took a personal interest in the reports of massive untapped oil reserves in the Canadian north. He saw the oil sands as a potential source of new profits, but also as a way to make the United States less reliant on foreign oil. "Something had to be done to ensure the continuity of North America's petroleum supply and prevent the continent from becoming increasingly reliant on crude imported from Venezuela or the Persian Gulf," he said. In the 1950s, Sun Oil made its first investment in the oil sands: $500,000 for research into how the company could profitably remove the bitumen from the ground and turn it into burnable oil.

If successful, Sun Oil would be doing God's work. The oil executive was a Presbyterian fundamentalist who believed the Bible was a literal description of historic events. He was convinced that "all of our so-called freedoms stem from Christian freedom. Without Christian freedom, no freedom is possible." The church, he argued, was "the only hope of the world." Pew saw creating a large oil industry in Canada as a way to counter the "godless communism" of the Soviet Union and its allies.

Pew's religious convictions would end up aiding Sun Oil in its negotiations with the Alberta government. In 1963, Dunlop testified before the Alberta Oil and Gas Conservation Board, where he assured regulators that oil sands development wouldn't interfere with the province's rapidly growing conventional oil industry. Premier Ernest Manning was enthusiastic about Sun Oil's ambitions but worried about the optics of an American company getting wealthy from Alberta's resources. Manning insisted that Sun Oil employ Albertans "as far as it is reasonable and practicable to do so," and also that it sell up to 15 percent of its initial equity stock to residents of the province. Executives at Sun Oil were not happy about these requirements, which they saw as "compromising [the] principles of free enterprise."

With tensions between Sun Oil and the Alberta government growing, Pew began exchanging private letters with Premier Manning, who also held deep evangelical beliefs and was a radio preacher. "Soon the correspondence assumed a tone of familiarity that was strengthened by talk about the Bible," according to an essay by Dochuk in the book *American Evangelicals and the 1960s*. "Pew told of upcoming conferences for Presbyterians and evangelicals, forwarded books for the premier's edification (*Calvin*, by Francois Wendel, and *The Man God Mastered*, by Jean Cadier), and asked if he could use Manning's sermons (no doubt in company settings)," Dochuk writes. The premier responded in one letter to Pew that "aroused Christian laymen" like themselves needed "to take an uncompromising stand for the faith."

Pew and Manning soon became good friends. In 1964, the two had a meeting on the banks of the Athabasca River where they shook hands and affirmed their "common commitment to Jesus Christ." Later that year, they met privately in Jasper, Alberta, amid the Rocky Mountains, to discuss their religious and business ambitions. Those conversations carried onto the golf course at one point, where in between rounds "Manning and Pew talked through theology and tar sands, and by all accounts the talks went well," according to Dochuk's account. Sun Oil received the premier's blessing to get an oil sands operation started as quickly as possible.

Manning had his own personal reasons to accelerate the extraction of bitumen. Tapping one of the world's biggest oil reserves would no doubt be great for Alberta's economy. But the premier's religious beliefs "were essential to how he looked at oil," explained Dochuk. "He saw oil as very much a gift from God." As a "premillennialist" who was convinced that Christ would be returning to Earth within his lifetime, Manning believed that the province only had limited years to develop its bitumen. "The sense [was] that time was short. Man, humanity needed to make the most of its resources in this current moment. To also foster the expansion of the Christian church and to spread the gospel," according to Dochuk.

With the cross-border political hurdles for Sun Oil cleared, the company now had to actually figure out the technology that would make developing the oil sands feasible. Around the time of Dunlop's visit to northern Alberta, Sun Oil announced it would spend $250 million on that challenge—at the time the largest single private investment in Canada. It was a huge spending commitment for an experimental and unproven oil operation. Adjusted for inflation, it would be equivalent to almost $2 billion in today's dollars. Longley, the PhD researcher, believes Pew's personal obsession with the oil sands, motivated by his deep religious aversion to communism, was a significant factor in making the investment possible. Sun Oil, a private company owned by the Pew family, "pumped a lot of money that might not have been so forthcoming in a public company," Longley said.

A breakthrough came in the mid-1960s, when Sun Oil figured out that the bitumen could be extracted with bucket-wheel excavators. These machines, ten stories tall, had first been used in Germany during the 1950s to extract coal, and were not common in North America. But they proved to be well suited to the extreme boreal forest conditions. Mounted on tank treads, the bucket wheel resembled a buzz saw on wheels, slicing into the earth. The bitumen was then deposited onto a conveyor belt and transported to a nearby processing plant, avoiding the need for trucks to navigate the muskeg. Even then challenges remained, however, particularly during winter months. "Buckets broke teeth on the frozen ground, large chunks of frozen bitumen jammed the crushers and the conveyor belts split in the extreme cold," an observer said at the time.

Inside the processing plant, the bitumen was mixed with hot water and corrosive soda, a process building off scientist Clark's 1920s experiments, creating a thick sludge that was then transported into a separation tank, where heavier waste sunk to the bottom. The oily substance remaining then entered a centrifuge, where sand, clay, and other impurities were removed, eventually resulting in smoothly flowing fuels that could be sent down a pipeline.

The Great Canadian Oil Sands site officially opened on September 30, 1967. The day was cold and overcast. But that didn't stop dozens of airplanes and jets from arriving at Fort McMurray's airport, "carrying oil executives, government officials and other dignitaries," the *Calgary Herald* reported. It was "a crowd of about 600 VIPs predicting incredible wealth would be unlocked from the massive energy resource in northern Alberta." Among them were emissaries from a rapidly expanding U.S. petro-empire, celebrating an important new outpost in northern Canada.

Premier Manning addressed his audience from a podium. "We are gathered here for this ceremony to officially open this gigantic complex, which for the first time will tap commercially the vast supply of oil which until now has remained locked in the silent depths of these Athabasca tar sands," he said. By Manning's side was Pew, who declared that "no nation can long be secure in this atomic age unless it be amply supplied with petroleum... oil from the Athabascan area must of necessity play an important role." With the assistance of a local worker, Pew pushed the buttons that got the operation's process lines moving.

"For them," Dochuk later said of the moment, "it seemed like the answer to a prayer."

In reality, as thousands of people in the Philippines would learn half a century later, it was a curse.

"We were all so desperate"

JOANNA SUSTENTO was starting to feel jittery. Not just for herself but for everyone huddled within the vibrating walls of the little bungalow in Tacloban City: her father, mother, two brothers, sister-in-law, and toddler nephew Tarin. "We were all looking outside, trees were already uprooted, our walls were destroyed, windows cracked, and then we heard a blasting sound," she recalled. Sustento's family didn't know it then, but Typhoon Haiyan had become one of the strongest storms in recorded history.

As more water rushed through cracks around the kitchen door, she and the others realized they were wrong to assume it was rain seeping in. The water was coming in from the sea—rapidly. Within only a few seconds, it was touching Sustento's knees. "So I went to my room and quickly grabbed my backpack and packed up my important things," she recalled. "As I went out of my room, the water was already chest level. We all decided to go out of the house because if not, we would be trapped inside."

Exiting the house was difficult because of the water rushing outside their front door. It was Nanay—the name Sustento called her mom—who helped pull her out. The family held onto window grills on the outside of the house and tried to stay calm. Suddenly Sustento's sister-in-law, Geraldine, screamed. She'd been bitten on the thumb by a snake. Her dad instructed Sustento to suck the venom and spit it out. "I wasn't sure of what I was about to do, but I did it anyway because we were all so desperate for survival," Sustento said. Her dad had recently had heart surgery and Sustento remembered worrying that if he became too panicked or if she disobeyed his instructions he might have a heart attack. With the water rising, her brother said they should all try to get on the roof.

Sustento's mom was able to climb up first. Next, they tried to get little Tarin up. But Geraldine was having a hard time carrying him because she felt weak. "Maybe it was because of the snake bite," Sustento said. They decided getting the whole family on the roof wouldn't work. But back on the ground, dangers were mounting. The wind felt like crushed ice and the waves were getting stronger. Sustento's brother Julius saw a life jacket floating by. He swam out and grabbed it, then returned and put it on Tarin. Then Julius saw a cooler bobbing by. Maybe they could use it as a flotation device, he thought. Julius swam out again and was able to push the cooler toward the family. But the currents prevented him from swimming back. That was the last they saw of Julius for the time being. The water was churning so hard that Sustento felt as though she was in a washing machine.

In the pandemonium, family members started to lose track of each other. Sustento saw Geraldine being pulled away in the water, holding on to a tree branch, and Tarin floating nearby in his life jacket. Sustento started screaming for Geraldine to grab Tarin. "When I saw her hand [reach] for Tarin I focused myself on saving my parents," Sustento said. Her mom was still up on the roof, and Sustento told her to come down and hold on to a log that she and her dad were using to stay above the water. "At that point, I could see the worry in their eyes that they were trying to hide," she recalls. Sustento's dad kept getting pulled under. "He was struggling to surface the water, gasping for air." Sustento turned her eyes toward her mom. There was terror in Nanay's eyes. "She just kept saying, 'Lord, save us!'"

A "hellish cloud"

AFTER THE POLITICIANS and oil executives and VIPs wrapped up their celebratory launch of the Great Canadian Oil Sands plant on that September day in 1967, the workers who remained had the task of actually keeping it running. It wasn't just Canadians who depended on their success, but shareholders of the American company Sun Oil. And behind them were the U.S. politicians and policymakers who saw the oil sands as a crucial oil supply from a friendly neighbor that could help America beat back the global spread of communism.

The plant was supposed to be producing 45,000 barrels of bitumen per day, but in those early months it could only manage a third of that, and even sustaining that level of production was difficult. A worker named George Skulsky remembers a night that winter when all of the powerhouse boilers failed. "It was 40 below outside and... the whole plant went down cold. Everything froze up," Skulsky later told the Canadian Press of a shutdown that lasted months. "It cost millions and millions of dollars. They had to replace a lot of equipment to get the plant back up and running."

In the meantime, the international oil and gas association that Sun Oil belonged to, the American Petroleum Institute, was receiving much more worrisome news: the fossil fuels it was spending hundreds of millions of dollars to extract from northern Alberta and elsewhere were destabilizing the Earth's climate. It was similar to the warning that Edward Teller had given Robert Dunlop and hundreds of others at Energy and Man a decade earlier. This time, though, the warning wasn't coming from an outside source. The climate science that Sun Oil and other oil companies were now becoming aware of was generated privately by the industry's own scientists.

The oil and gas industry by that point had more than two decades' experience studying air pollution from its products. Its first official research initiatives began in the 1940s as a reaction to a growing economic threat in Los Angeles. A suburban boom was underway at the time, and as millions of cars clogged the California city's new freeways, the air was getting dangerous for its residents to breathe. On one day in 1943, the smog got so bad that people could only see three blocks ahead of them; many assumed that it was a chemical warfare attack by the Japanese army. Throughout the decade, Los Angeles regularly experienced air pollution that turned the sky a sickly yellow and caused people's eyes to burn if they spent too long outside. One government report referred to the smog as a "hellish cloud."

The city's very identity was at stake. "People in Los Angeles were very proud of their air... They said that L.A. was the land of pure air, and that moving there could cure tuberculosis and alcoholism," Chip Jacobs, one of the authors of *Smogtown: The Lung-Burning History of Pollution in Los Angeles*, explained. At first, the origins of the smog were unclear. Many people figured it was being caused by the Southern California Gas Company's Aliso Street plant, but even after public pressure resulted in that plant being temporarily shut down, the city's air continued to deteriorate. Others pointed to a different culprit: the oil and gas industry.

Around that time, executives from the Western Oil and Gas Association, a trade group representing major fossil fuel companies operating on or around the West Coast, created the "Committee on Smoke and Fumes." Its mission was to fund and conduct industry research on the causes of Southern California's smog and then disseminate that research to journalists and policymakers. The goal, as a document from the American Petroleum Institute explained, was "to avert restrictive and uneconomic [regulations] of the type that had proved unnecessary in the past." From its earliest years onward, Smoke and Fumes had no problem suppressing science that could harm the business model of its members, as the story of Harold Johnston makes clear.

In 1950, Johnston was a young graduate student finishing his thesis at Stanford University. One day he attended a seminar featuring the scientist and air pollution researcher Arie Haagen-Smit. The scientist, Johnston later recalled, was "Dutch, a very precise man, and a bit astringent to deal with." But Haagen-Smit was a brilliant researcher, and after studying Los Angeles's smog, he had come to the conclusion that it was the result of gasoline being burned in car engines, not the industrial factories in and around the city that many people still assumed were the cause. "People were thinking smog came from factory smoke," Johnston said.

As Haagen-Smit explained his calculations during the seminar, Johnston thought he detected an error. During a question period following the scientist's presentation, Johnston "stood up at the meeting and politely pointed out that Haagen-Smit's first reaction [in his theory] was too slow to produce the observed build-up of ozone every day [in the city]." Watching this academic exchange from the back of the room was an oil and gas industry representative named Vance Jenkins. After the seminar, Jenkins traveled to the Stanford Research Institute, a new science lab that received a large portion of its funding from the oil and gas companies behind the Smoke and Fumes Committee. Jenkins recommended Johnston, the questioning graduate student, for a job. "[He] said in effect, 'Hey this brash young guy stood up and said Haagen-Smit's mechanism is wrong. Why don't you hire him and find out what he's talking about,'" Johnston recalled.

Johnston was shortly after brought in for an interview with the Stanford Research Institute's board of directors. "Especially the oil people were vociferous that Haagen-Smit was all wrong, that he was only a publicity-seeker. Automobiles and gasoline couldn't have anything to do with [smog]. They really opposed him on a personal level," he said. The thing was, Johnston explained to the directors, he didn't really know that much about Haagen-Smit's theory linking the burning of fossil fuels to smog—his objection had only really been an off-the-cuff reaction during the seminar. But his interviewers were undeterred. "They said, 'We'll pay you for studying the literature here and coming back to explain what you found.' They said terrible things about Haagen-Smit," Johnston said. "So I was given the job of disproving the theory of this quirky, perhaps dishonest, scientist."

Johnston went to work immediately, poring through Haagen-Smit's articles and presentations. "I rapidly concluded that Haagen-Smit was a genius!" he said. "Everything he said was based on experiments he had carried out, and his techniques seemed to be sound." Johnston presented this new information to a manager at the Stanford Research Institute. He didn't get the response he was expecting. "The manager said, 'Oh, this is important, this is very important, but we must be careful. We have to do more research before we can bring this out,'" Johnston said. A presentation Johnston was supposed to do for the institute's board of directors was promptly canceled, and not long after the young scientist's position was terminated.

Years later, the Smoke and Fumes Committee was still arguing that Haagen-Smit's scientific work linking the combustion of gasoline in car engines to Los Angeles's smog was incorrect. "The work at Stanford Research Institute has shown that there are a number of apparent errors both in this theory and in its interpretation to account for the various phenomena associated with smog," Vance Jenkins wrote in a 1954 publication, disregarding what Johnston had earlier discovered proving the smog theory correct. Jenkins added that if governments across California and other U.S. states passed legislation to crack down on emissions from gasoline and oil refineries, it "would result in large and unnecessary expenditures by the oil industry."

Smog was just the beginning, however. By the 1960s, new science linking global temperature rise to the burning of fossil fuels was beginning to be recognized at the highest levels of the U.S. government. "Air pollution is no longer confined to isolated places. This generation has altered the composition of the atmosphere on a global scale through radioactive materials and a steady increase in carbon dioxide through the burning of fossil fuels," president Lyndon B. Johnson said in 1965. That statement, the first of its kind uttered by a U.S. president, followed a report from Johnson's science advisers that contained a chapter entitled "Carbon Dioxide From Fossil Fuels—The Invisible Pollutant."

The scientists warned the Democratic president that at the current levels of burning oil, gas, and coal, concentrations of carbon dioxide in the atmosphere could grow 25 percent by the year 2000—this, they predicted, could cause "melting of the Antarctic Ice Cap," "rise of sea level," and "warming of sea water," and could "increase acidity of fresh waters." "Ours is a nation of affluence," Johnson wrote in the report's introduction. "But the technology that has permitted our affluence spews out vast quantities of wastes and spent products that pollute our air, poison our waters and even impair our ability to feed ourselves." The president recommended that federal agencies "report to me on the ways in which we can move to cope with problems cited in the report."

The oil industry was not pleased. "This report unquestionably will fan emotions, raise fears and bring demands for action," American Petroleum Institute president Frank Ikard said during a speech at an annual meeting of the organization later that year. "The substance of the report is that there is still time to save the world's peoples from the catastrophic consequence of pollution, but time is running out." He went on, "One of the most important predictions of the report is that carbon dioxide is being added to the earth's atmosphere by the burning of coal, oil and natural gas at such a rate that by the year 2000 the heat balance will be so modified as possible to cause marked changes in climate beyond local or even national efforts."

Robert Dunlop, who was at that point dedicating much of his time to getting Canada's first oil sands operation started, appears to have been at this meeting. At one point Ikard refers directly to the Sun Oil president's attempt to put a positive spin on the industry's environmental challenges. "As Mr. Dunlop said," Ikard reports, "I think [those challenges are] really opportunity."

Not long after, just like it had done with smog two decades earlier, the American Petroleum Institute turned to the Stanford Research Institute, asking two of its scientists, Elmer Robinson and R. C. Robbins, to study global temperature rise and its link to the oil and gas industry.

The duo delivered their report in 1968. In it, the scientists concluded that the earlier climate research done for President Johnson was largely correct—if anything, it was too conservative. Whereas the Johnson report predicted that concentrations of carbon dioxide could reach 350 parts per million by the end of the century, the petroleum scientists thought it could be as high as 400 parts per million. (The actual figure ended up being 370 ppm.) At this level of warming, Robinson and Robbins warned, "significant temperature changes are almost certain to occur." They went on: "There seems to be no doubt that the potential damage to our environment could be severe."

The scientists considered several explanations for the rapid global warming taking place, but "none seems to fit the presently observed situation as well as the fossil fuel emanation theory"—that is, that carbon dioxide emissions come from burning the oil industry's products. Given all their worrying findings, Robinson and Robbins recommended further research "toward systems in which CO_2 emissions would be brought under control."

Did the advance knowledge that the oil and gas industry's products could permanently alter the atmosphere and flood many of the world's coastal cities affect Sun Oil's plans for the oil sands? Apparently not. In 1968, the year scientists commissioned by the

American Petroleum Institute warned of "severe" impacts from climate change, Sun Oil began sending the very first oil from its Great Canadian Oil Sands plant down a pipeline from Fort McMurray, Alberta, to Sarnia, Ontario.

Those early barrels would lead the way for billions more to come.

"Operation Oilsands"

WITH SUCH HUGE AMOUNTS of oil buried in northern Alberta, it was only going to be a matter of time before the multinational oil company Exxon got involved. By the 1970s, Exxon was the second-largest corporation in the United States, with oil and gas operations on every continent except Antarctica. Yet even with its vast financial resources and political connections, it nearly failed in building a bitumen operation that could compete with the one created in 1967 by Sun Oil. The journey that led Exxon to become the second major oil sands producer required hundreds of millions of dollars, 450,000 tons of construction materials, over six thousand acres of strip-mined boreal forest, enough steel pipe to stretch from New York to Washington, D.C., and a reckless development plan that nearly turned large parts of the region radioactive.

Exxon has long been one of the world's most iconic—and infamous—companies. But in Canada it operates under the much less recognizable brand name Imperial Oil. The first iteration of Imperial Oil was Canadian; it was started in 1880 by sixteen oil refiners from Ontario to tap and refine the country's barely touched oil reserves. Eight years later, however, the company became American controlled after selling off a majority interest to Standard Oil, the New York–based petroleum behemoth that eventually spawned

Exxon. To this day, Exxon owns nearly 70 percent of Imperial Oil's shares, making the ostensibly Canadian company a vector for U.S. interests.

Not long after being bought out by America's largest oil producer, Imperial began searching for ways to exploit the oil sands. The company purchased seventeen production leases in 1917, following a survey of the region conducted by several American geologists. But like most others that had seen vast potential around Fort McMurray, the company concluded that actually transforming the thick bitumen into fuel that could be sent down a pipeline and pumped into a car engine was going to be difficult. "If there was oil in the Athabasca region," an investigation it conducted at the time concluded, "it was not going to yield to traditional methods of drilling." Several decades went by without much action from Imperial. But in the 1950s the company thought it had found a solution. The result, historian Graham Taylor explains, was "one of the most bizarre episodes in the history of the oil sands."

During that time period, the U.S. government was experimenting with using nuclear weapons for peaceful purposes. The program, dubbed Project Plowshare, was officially set up under the Dwight Eisenhower administration in 1958, its name a reference to a biblical injunction urging people to "beat their swords into plowshares, and their spears into pruning hook; nation shall not lift up sword against nation, neither shall they learn war any more." The idea behind the project was that the United States could use controlled nuclear explosions to build canals, blast through California mountains for railroads and highways, enlarge harbors in Alaska, or accomplish anything else requiring large excavations. One of the project's most enthusiastic backers was Edward Teller, the nuclear scientist who warned Robert Dunlop about climate change in New York.

The program also piqued the interest of Manley Natland, a petroleum geologist working for the California company Richfield Oil. What if, Natland wondered, nuclear bombs could be used to tap the

Canadian oil sands? This idea apparently came to him during a trip to Saudi Arabia's southern desert in the 1950s, as he later recounted in a paper: "One evening as I sat watching a spectacular sunset from a small hill overlooking a flat, endless sea of sand, the sun looked like a huge orange-red fireball sinking gradually into the earth," he wrote. Crazy as it sounds now, "Natland's unusual sunset-inspired vision generated great interest in the Alberta capital," historian David Breen recounts.

Knowing Imperial Oil had leases in the oil sands, Richfield reached out and asked if the company wanted to come in on a plan to set off a nuclear explosion in the region. Imperial agreed that this was a promising idea, and used its connections with the U.S. and Canadian governments to set up presentations for the proposal at the U.S. Atomic Energy Commission, the Atomic Energy Control Board of Canada, and the Canadian federal Department of Mines. In 1958, Richfield representatives traveled to Edmonton, Alberta, where they outlined to policymakers an "experiment in the peaceful use of nuclear energy as an aid in producing oil from the McMurray oil sands buried too deeply to permit economic extraction of oil by mining methods."

The plan called for drilling a well about sixty miles south of where Sun Oil would later build its own plant, and then detonating a nine-kiloton nuclear device. "It was expected that the resulting explosion would create a 230-foot-wide cavity, into which would drain several million cubic feet of oil released by the explosion's tremendous heat," Breen explained. "Also, it was claimed that there would be no radioactive fallout since the explosion would be contained underground."

Natland was adamant it could be done safely. "[You] would go into the centre of the township and seed an area with all these heat cells [created by the nuclear explosion] and then perhaps let it sit for a year. That would keep the heat completely spread around, and do as much good as it possibly could in fluidizing the oil and then we could come back later and [remove the oil by conventional means]." It would give off about as much radiation as an average wristwatch, he claimed.

In February of 1959, the Alberta government, along with Canadian officials and representatives of the U.S. Atomic Energy Commission, did a press conference about "Project Oilsands," whose original name, "Project Cauldron," had been scrapped following fears that the public would find it too "effervescent." A front-page story in the *Calgary Herald* newspaper proclaimed that the plan "will give the Western world a measure of independence from huge Middle East oil deposits." But not everyone shared that enthusiasm. One provincial official warned that "if it does not turn the whole deposit into a burning inferno, it is absolutely sure to fuse it into a solid mass of semi-glass or coke." A nuclear scientist in Utah predicted such an underground explosion could "spread radioactive dust for more than 200 miles."

A health official named Dr. D. Dick was also worried. He agreed that the health risks were too great. But facing internal pressure from government supporters of the project, he wrote in a committee report that "setting off a nuclear device beneath the Oil Sands in Northern Alberta should pose little or no hazard to health or safety." However, Dick added the disclaimer that "what constitutes a safe level of radiation is not yet known." Dick wasn't the only one worried. Many Alberta residents wrote letters to premier Ernest Manning expressing concerns. Growing public unease about nuclear safety, combined with increasing skepticism from the federal government about the wisdom of placing U.S. nuclear weapons in Canada, led to "Operation Oilsands" being "indefinitely postponed" in the autumn of 1959.

But Imperial Oil and its partners decided to proceed in the oil sands anyway. With the nuclear plan on hold, Imperial, Richfield, and another partner officially embarked on a project about 310 miles north of Edmonton at Mildred Lake using conventional mining techniques. They named their consortium "Syncrude," a reference to the "synthetic crude" produced from the oil sands. The boggy muskeg and freezing winter presented similar challenges to those Sun Oil had also experienced during the 1960s, resulting in years of setbacks, delays, and mounting costs. By 1968, the projected costs were $800 million, a massive investment in those days. Richfield soon fled from the project. Large new oil discoveries in Alaska's Prudhoe Bay around the same time raised broader doubts about the oil sands' financial viability, and by 1974, due in part to inflation, estimated costs for the project surpassed $2 billion. Meanwhile, Imperial's own scientists were raising internal concerns about climate change.

One of the first warnings came from a chemical engineer named H. R. Holland, who was employed in Imperial Oil's engineering division. The warning was contained in a 1970 report he wrote entitled "Pollution Is Everybody's Business," whose contents included a patronizing history of colonialism in North America, along with surprisingly candid reflections on actions necessary to curb the oil and gas industry's harmful emissions.

"Consider the fate of the buffalo herds of the great plains," Holland wrote. "To these herds, the Indians were a foreign and hostile addition to their environment. Fortunately, the number of Indians was small and their arms were primitive so that the herds could adjust to their presence. Then the white men came with repeating rifles and the degree of interference became intolerable. The environment of the buffalo was polluted, and the great herds vanished. Typically the results of this pollution were varied: for the buffalo—extinction; for the Indians—the end of their nomadic way of life; for the white men—the plains became available for agriculture and settlement."

He went on, "Since pollution means disaster to the affected species, the only satisfactory course of action is to prevent it—to maintain the addition of foreign matter at such levels that it can be diluted, assimilated or destroyed by natural processes—to protect man's environment from man." Holland argued that this task couldn't be left up to voluntary pledges from individual companies. "The protection of the interests of society as a whole requires the establishment of legal controls on pollution as on other anti-social acts," he wrote.

Holland had in his report a list of "air pollutants" Imperial should be paying attention to. Along with smog-causing emissions like sulfur dioxide and nitrogen dioxide, the list included climate-warming carbon dioxide. This is important, a U.S. fossil fuel watchdog group called the Climate Investigations Center later explained, because it shows that the company "was aware as early as the late 1960s that global emissions of CO_2 from combustion was a chief pollution concern affecting global ecology."

By the mid-1970s, it was unclear if Imperial Oil's oil sands project at Mildred Lake would actually proceed. With Richfield gone, costs continued to rise. When negotiators for the Syncrude project met with the Alberta government in February 1974, the project was so expensive that "the remaining partners could realistically threaten to close it down," Taylor explained. Federal and provincial policymakers were eager to keep the project going, though, due to high oil prices and shortages caused by Middle Eastern producers restricting exports during the OPEC crisis. The Canadian government ended up taking a 15 percent ownership stake in Syncrude. Alberta's government acquired an additional 10 percent, while extending a $200 million loan it had earlier made to the Imperial-led project. "The dream was saved and construction hit fever pitch," an official Syncrude history explains.

But as before, money wasn't the only concern being raised inside the company. In 1977, Imperial's parent company Exxon received one of the industry's strongest climate warnings yet. During a presentation at corporate headquarters in Irving, Texas, a senior company scientist named James Black briefed executives about the current state of climate science. “In the first place, there is general scientific agreement that the most likely manner in which mankind is influencing the global climate is through carbon dioxide release from the burning of fossil fuels,” Black told Exxon’s management committee, according to documents acquired by the media outlet *Inside Climate News*. If those emissions didn’t slow down, he warned, the dangers to humankind could be immense.

Black sketched out what could happen during a presentation to Exxon scientists and managers the following year: global temperatures could eventually increase by two to three degrees Celsius (about 3°F–5°F), fundamentally altering the conditions for life on Earth. “A possible result might be a shift of both the desert and the fertile areas of the globe toward higher latitudes. Some countries would benefit but others could have their agricultural output reduced or destroyed. The picture is too unclear to predict which countries might be affected favorably or unfavorably,” he said, according to a written summary of his remarks. “Even those nations which are favored, however, would be damaged for a while since their agricultural and industrial patterns have been established on the basis of the present climate.”

Black told the company that it needed to act quickly on this knowledge. “Present thinking holds that man has a time window of five to ten years before the need for hard decisions regarding changes in energy strategies might become critical,” he said in a document addressed to F. G. Turpin, vice president for Exxon Research and Engineering.

That document was sent on June 6, 1978. Less than two months later, the Syncrude project at Mildred Lake began pumping its first oil, roughly fifty thousand barrels per day. The plant officially opened on September 15. It encompassed an area of almost ten square miles, equivalent in size to a small town. The final cost after twenty years of planning and five years of construction was $2.2 billion. There were six hundred guests there that day to celebrate the opening, including federal leaders and oil executives representing interests on both sides of the U.S.-Canada border. "[They] bumped around the massive 27-acre site by bus," *Maclean's* magazine reported from the scene. "At the ceremony, speaker after speaker proclaimed such plants to be Canada's future energy hope."

Nobody mentioned climate change.

"A million pieces"

AS JOANNA SUSTENTO and her mother fought the pull of the rising current, staring into each other's petrified eyes, a refrigerator floated by, and they were able to grab hold of it. By now they'd drifted up against the side of a building behind their house, where wave after wave crashed into them. "I was afraid to be trapped so I pushed the refrigerator away and held on to a log," Sustento said. "That was the time that Nanay and I got separated." It was getting harder for Sustento to keep her head above water. She'd be forced under and then pop up gasping for air. She was exhausted and unsure how much longer she could keep going. Then a wave pulled her down, and when she tried to surface she found her head blocked by debris. She was running out of breath and there seemed to be no way out. "I just gave up and said, 'If I'm going to die today, then so be it.' I asked for forgiveness, I was praying and saying 'I love you' in my head to everyone I love."

Similar scenes of chaos were taking place all over Tacloban. A woman and her cousin climbed over a wall and took shelter on top of a neighbor's house. Another woman was swept into the waves with her five-year-old son clinging to her back and held onto a coconut tree for hours. A tourist from North Yorkshire took shelter in a diving shop and saw metal, wood, and glass debris whooshing by in the air. A young mother who couldn't swim asked her husband to save their daughter and ended up surviving. Another man lost twenty family members. A storm surge reaching heights of twenty feet swept across the city, completely submerging the surrounding buildings. The waves were so strong that they pulled a cargo ship, the M/V *Eva Jocelyn*, into a low-income coastal area known as Barangay Anibong, destroying concrete homes and crushing several people.

Sustento was certain she would drown when what seemed like a miracle happened. Just as she stopped thrashing her limbs and slipped underwater, some of the debris above her head shifted. "The weird part was, the moment I decided to let go and accept what was happening, I was able to surface," she said. Sustento saw people on a neighboring roof waving to her and shouting. She looked across the water and saw her mom calling out for help. "She sounded so desperate. And it broke my heart to see her that way." Sustento grabbed a piece of wood and floated over to her mom, who was being buffeted by waves. "I tried helping her to lift half of her body, but when my hands slipped down from her arms to her fingers, her body splashed into the water, and that was when I realized Nanay was gone," Sustento said. "She didn't struggle anymore. The look on her face broke not just my heart, but my whole being was smashed into a million pieces."

Waves kept hitting them, pulling her mom's body off the piece of wood. A current carried them toward a water tank with steel frames. "So I locked my left leg onto the frame as I was trying to lift Nanay with both my arms and my right leg," she said. "It was difficult to bear seeing Nanay helpless, lifeless; her beautiful smiling face was now a face in shock." Sustento had to make one of the hardest decisions of her life. "Am I going to hold on to Nanay and die as well? Or should I let her go and save myself?" she said. "I seriously considered just dying with Nanay at that point. I didn't want other people to think that I didn't do anything to save her, to save them. I didn't want others to blame me because they died and I lived. And what good is my life now that my family was taken away from me?"

But then she began wondering who of her other family members might possibly make it out alive. And who might not. "What if Tarin survived? What if his parents didn't? Who would take care of him? Who would feed him? I wanted to survive for them. More importantly, for Tarin. I loved him dearly, as if he was my own."

Sustento knew she had to say goodbye to her mom.

"Staring at Nanay's face, I asked for her forgiveness, for everything I did that hurt her. I thanked her for everything she has done for me, for us. I thanked her for being the strong woman that she is, that even during her last moments in this world, she still managed to look out for me and save me. I embraced her so tightly and kissed her for the last time."

Then Sustento let go.

II

The Early Construction of Denial

(1968–1988)

“He seemed embarrassed”

AS CHARLES KOCH flew over the Alaskan wilderness in his corporate plane, he debated with two other associates what to name the corporate empire he’d inherited. The year was 1968, and Koch, who was in his early thirties, had just been given control along with his brothers of a series of fossil-fuel-heavy ventures created by his father Fred Koch, the strict family patriarch, who’d died suddenly of a heart attack the year before. Charles had quickly gone to work helping consolidate all these businesses into a single company. But what to call it?

With Alaska’s snowcapped mountains passing below as they headed to the company’s headquarters in Wichita, Kansas, the plane’s travelers debated one promising suggestion: Koch Industries. Charles wasn’t too keen on it at first, Christopher Leonard recounts in *Kochland: The Secret History of Koch Industries and Corporate Power in America*. “He seemed embarrassed by the thought of having his last name stamped on the entire company. His name would be embossed on the letterhead, emblazoned on the sign outside the company headquarters, spoken on the lips of everyone who worked for him,” Leonard writes. Charles was known within the Koch family for being fiercely competitive and self-confident. But he also had the publicity-avoiding temperament of someone who immersed himself in complex mathematical equations and abstract political theory, having graduated with master’s degrees in nuclear and chemical engineering from MIT and at one point during his post-college years living "almost like a hermit, surrounded by books," as Charles would later recall to the *Wall Street Journal*.

But Roger Williams, who'd recently been hired to run the company's sprawling network of oil pipelines, made the case in favor of having a company name that was neutral and difficult to pronounce. Public-facing companies with a consumer product to sell such as Coca-Cola required a catchy name that stuck in people's minds. "But the oil industry was different because Big Oil was cast as the villain in so many economic stories," Leonard writes. "For this reason 'Koch' was the perfect moniker for the firm. It was slippery, hard to grasp." Charles was apparently won over by Williams's arguments. Not long after the flight from Alaska, he made the company's new identity public.

The company was at the time bringing in estimated annual revenues of around $177 million. But a business decision made by Charles during his early years at the corporate helm laid the foundation for Koch Industries to become one of the wealthiest and most powerful private companies in the world, while opening up the floodgates of oil sands oil into the United States, which in turn funded a torrent of reactionary politics. Charles would later describe it as "one of the most significant events in the evolution of our company."

Among the assets that Charles and his brothers had inherited from their father, Fred Koch, was a minority stake in the Great Northern oil refinery, a processing facility for petroleum outside of Minneapolis. The refinery was situated near the Pine Bend Bluffs nature reserve, a sprawling expanse of savanna, oak forest, and prairie overlooking the Mississippi River.

In 1969, Charles Koch convinced the other two owners to sell their shares in the refinery, making Koch Industries the sole owner. At the time, it wouldn't have seemed obvious why this was a great deal for Koch Industries. New refineries were being built frequently in the United States in those days, and competition was tough. But in Koch's eyes the Minnesota project, later named the Pine Bend Refinery, had huge advantages. Because the refinery was so close to the Canadian border, it could take advantage of a loophole in federal energy policy. There were strict limits dating back to the 1950s on the amount of oil that U.S. refineries could import from other countries. Crude imports from Canada, which was seen as a trusted ally, were unrestricted.

Not that many refineries in the Midwest had taken this opportunity, because crude from Canada tended to be lower quality than the smooth-flowing conventional oil pumped from places like Texas, and thus required expensive specialized equipment. But the refinery Charles Koch took over had specialized coking equipment already installed, allowing it to purchase Canadian crude that was selling at bargain prices due to a relative lack of buyers. (The company would later refer to these oil sources in legal filings as "garbage crudes.") Pine Bend was also one of the only major refineries in the Midwest, a fact it took advantage of by selling refined products at a large markup. Revenues from the refinery were so immense and consistent that decades later executives at Koch Industries would still be referring to it as the company's "cash cow."

In addition to the stake in Pine Bend, Charles had inherited the far-right politics of his father, a strident anti-communist who'd cofounded the John Birch Society, a political advocacy group that argued during the 1960s that the civil rights movement was created by the Soviet Union in order to dismantle capitalism in the southeastern United States and create a "Soviet Negro Republic." Charles apparently rolled his eyes at the John Birch Society's wilder conspiracy theories and considered a significant portion of it to be "bullshit," according to an acquaintance of the family.

But the group's basic worldview—that any attempt by the government to redistribute wealth by assisting the poor, regulating industry, or taxing wealthy individuals is fundamentally illegitimate—proved appealing to Charles. Prior to taking over his father's fossil fuel empire, Charles became an executive and trustee at the Freedom School, a Colorado Springs–based organization that "taught a revisionist version of American history in which the robber barons were heroes, not villains, and the Gilded Age was the country's golden era," according to Jane Mayer in *Dark Money: The Hidden History of the Billionaires Behind the Rise of the Radical Right.*

As the profits of Koch Industries grew exponentially, Charles became ever more deeply involved in the far-right political ecosystem his father had helped create through John Birch. Four years after gaining majority control over Pine Bend, Charles took over a think tank called the Institute for Humane Studies, which deemed all taxes collected by the government "theft." The young oil executive and his brother David Koch cofounded their own think tank in 1977 that would later be known as the Cato Institute. The new think tank's ideology could be summed up in an essay Charles wrote that same year in a publication known as the *Libertarian Review*, in which he claimed that any government intervention in the economy "is not only inefficient, it is thoroughly immoral as well... To return to that system is to finally abandon the American experiment and the American dream."

Charles apparently failed to appreciate the irony that the Pine Bend Refinery's early profits were arguably only possible due to the Eisenhower government's 1950s restrictions on oil imports. Those profits were later cemented by increasingly strict federal environmental regulations that made the construction of competing refineries in the Minnesota area prohibitively expensive.

Whether it was conscious or not, Charles's relentless fears of encroaching socialist government combined with his ability to see profit-making potential in unconventional sources of oil that others considered marginal made him a natural ally of the fledgling oil sands industry in Canada. During the same years that Charles was consolidating control over Pine Bend, Sun Oil was opening the industry's very first commercial operation in the muddy boreal forest muskeg near Fort McMurray. Like the Kochs, Sun Oil founder and oil sands pioneer J. Howard Pew was a staunch anti-communist who'd spent decades funding and supporting the far right, arguing that his mission was "to acquaint the American people with the values of a free market, the dangers of inflation, the paralyzing effects of government controls on the lives and activities of people."

Starting in the late 1960s, Koch Industries began buying up large landholdings in bitumen-rich areas of Alberta, according to research from the media outlet *Inside Climate News*. Regulatory filings suggest that at one point, a Koch Industries subsidiary known as Cold Lake Great Northern was "the sixth largest land holder in the Cold Lake oil sands area." By 1983, the company held leases on approximately sixty thousand acres in northern Alberta. "Everyone knew the Cold Lake acreage was rich in oil sands," the filings state.

Koch Industries never became a major producer of tar sands crude on par with Pew's Sun Oil or the Exxon subsidiary Imperial Oil, but the company got involved with the industry in other crucial ways. Beginning in the early 1980s, Koch Industries became among the first U.S. refineries to begin processing oil sands crude. That was the period during which Koch Industries did its first test runs at its Pine Bend Refinery of bitumen shipped southeast via pipeline from Sun Oil's mine near Fort McMurray, Alberta.

Like the Kochs' far-right politics, the early bitumen they processed was far outside the mainstream. "Pine Bend had temporarily been running an unusual crude (called Great Canadian Oil Sands, or 'GCOS'), which was a distressed product from Canada that was only partially refined," according to notes from a 1982 board of directors meeting for Koch Industries. "The GCOS was inexpensive, but Pine Bend was apparently at its capacity in de-sulfurizing this unusual feedstock." The rewards for figuring out how to consistently process the unusual oil source were potentially high: "Pine Bend could run more of [the bitumen] through the crude units than an ordinary crude." Being able to process more crude units could mean the refinery's already high profits would soar.

Charles's decision to gamble on obscure sources of oil from Canada was either incredibly astute or lucky or both. By 1981, the Pine Bend Refinery was far and away the most profitable venture in Koch Industries' thirty-two major divisions, bringing in $60.9 million—or 22 percent—of the company's $273.6 million in after-tax profits that year. The following year, the refinery's profits were up to $107.8 million, representing one-third of what the company cleared. "The refinery played a pivotal role in making Koch Industries one of the largest and most profitable companies in the world," Leonard writes in *Kochland*.

The president of Koch Refining Company, Bernard Paulson, told Leonard that Pine Bend was "the cash cow, really, that provided the early money for Charles to expand in other areas."

"Ahead of the game"

INDIGENOUS PEOPLES living in the northern Alberta community of Fort McKay watched with trepidation as surveyors from Sun Oil, Imperial, Koch Industries, and other U.S. oil companies traversed the oil sands region. Some Elders made dystopian predictions about what the area would be like once those companies began tapping into the vast bitumen reserves on a commercial scale.

Celina Harpe of the Fort McKay First Nation—which includes Cree and Dene speakers, along with the Métis descendants of Chipewyan people who'd intermarried with French and Scottish fur traders during the nineteenth century—was just a child when she sat beside the Athabasca River with her grandfather. "Look at the beautiful river, the way it looks now," her grandfather said on a spring day sometime in the years following World War II. "I see it, what's going to happen in the future. All the trees will be gone. They're going to dig big holes, and they're going to dig up all that black stuff. You know that tar? That's what they're after... I won't see it. I'm too old... But if you have children, you're going to have to tell them not to drink this river water."

By the time Celina Harpe was grown up, predictions like that were coming true. After the first oil sands plants opened upriver, many people in Fort McKay became convinced that the Athabasca was no longer a safe water supply. The community responded to those fears by getting two pump towers installed, which provided cleaner water driven in by truck from Fort McMurray. But in the early 1980s, one of the pump towers burned down in a fire. During the following winter, which was especially cold, the other pump tower froze. Community members responded by doing what they had previously done for thousands of years: going down to the river. For several weeks they took their drinking water from the Athabasca.

The timing couldn't have been worse. Right when all this was happening, a pond containing toxic waste at a bitumen operation upriver began leaking, sending contaminants seventeen times the legal limit into the Athabasca. The operation was run by Suncor, which had changed its name from Sun Oil several years earlier. Celina's sister Dorothy McDonald-Hyde was the community's chief at the time. Suncor apparently tried to phone Dorothy to warn her about the situation, but she wasn't available for the call. Suncor then waited three weeks to send representatives in person. All that time, community members were drinking potentially toxic water. Even though nobody had gotten seriously ill, the community was horrified.

When authorities in the Alberta government declined to take action against Suncor, the Fort McKay chief took the oil sands producer to court on her own—and won. The company was found guilty of polluting the Athabasca, but only had to pay an $8,000 fine. "Suncor ended up with a slap on the wrist, but [the lawsuit] was important because it showed people in Fort McKay that [they] could have an impact on the goings on in industry. It showed government and industry that they [Fort McKay] were going to follow through and not take things lying down," Dorothy later explained to *APTN News*.

Though the immediate penalty in this case was minimal, high-profile defeats like this carried a potentially serious long-term burden for oil sands companies. In public, bitumen producers rarely took seriously the concerns of Indigenous peoples. "At that time, Fort McKay was looked at as kind of being in the way," Dorothy explained. "They didn't consult with the community." But in private, those same producers saw First Nations' advocacy as a growing financial liability.

The true feelings of oil companies toward Indigenous peoples are hinted at in a series of documents commissioned by Imperial Oil from the late 1960s up into the 1980s. Those documents attempt to measure and anticipate the public's views on fossil fuel development. "In general, the attitude of Canadians toward the environment continues to exhibit a fairly high degree of concern," reads one document from 1973, which was also shared with representatives from Imperial's parent company, Exxon. The document warned that one result of this ecological anxiety might be that "new energy source developments with all their inherent environmental hazards could be stopped or, at worst, slowed down."

Further down in the section of the report labeled "public attitude," the document explained that "several of the major energy resources developments—oil, gas and electricity—include the problem of native peoples' aboriginal or treaty rights as well as environmental concerns." The problem, as Imperial saw it, was that these social issues "have become widely recognized" in Canada.

Imperial Oil's fear seemed to be that negative publicity generated by First Nations suffering at the hands of oil and gas companies could ripple out into the general public, causing a swing in public opinion that would compel policymakers to enact onerous and profit-killing regulations on pollution. Another internal document from this time period made the connection explicit: concern for the rights of "native groups," Imperial Oil explained, could be translated into policies that push "land-use planning instead of pure market-determined patterns of development," along with a "strong emphasis on reduced/managed energy demand and environmental protection." It was all part of a worrying trend: "the increasing 'politicization' of the petroleum business environment."

As far back as the late 1960s, Imperial Oil had begun to realize that public concern about the pollution its products caused was a growing corporate risk. A 1967 "Public Relations Assessment" for the company, which was labeled "confidential," provided some sense of the scale of the challenge. "The petroleum industry, directly and indirectly, is a major contributor to many of the key forms of pollution," it explained. "Unfortunately, air pollution is an area highly charged with emotion and one characterized by a lack of meaningful data and rational guidelines." The document went on to note that "the way the public, the mass media and governments react to these effects [of pollution], real or imagined, will have a profound impact on the future of the oil and chemical industries."

Imperial Oil cited 1967 public opinion research from the United States that showed 45 percent of people considered air pollution a major problem—an order of magnitude larger than just a few years earlier. The company in part blamed these "difficult-to-change anti-oil industry attitudes" on biased journalism. "Due to continuing exposure to stories in the mass media, the general public could easily be persuaded to support increased pollution regulation and legislation," the document warned. Even worse, the public "could be encouraged to support the electric car, nuclear energy and other technology favouring competitive fuels."

But Imperial Oil also saw civil society as a hostile influence on public opinion. During the 1970s, Imperial Oil began actively monitoring "public pressure groups" that it saw as contributing to political momentum behind less polluting alternatives to oil. Though the six organizations it compiled dossiers on differed in outlook and strategy, Imperial noted that they broadly shared a conviction that "government policy must provide for development of alternative sources of renewable energy and place greater emphasis on energy conservation." The document included a frank assessment of the strengths and weaknesses of each organization. A section on a public-interest group called the Canadian Arctic Resources Committee explained that the group's power lies in "its ability to induce reputable persons to offer critical observations about northern development. The high academic quality and integrity of these individuals lends considerable legitimacy to their arguments and makes an impact on the media."

Exxon was noticing a similar problem south of the border: antagonistic environmental activists had the potential to cause people to think negatively about the oil industry, resulting in costly regulations. Exxon predicted that concerns about air and water pollution were just the start. Debates on these issues, company leaders were convinced, would soon evolve into protracted public fights over climate change. "Atmospheric Science will be of critical importance to Exxon in the next decade," the company wrote in a 1979 document anticipating future legislative battles. Exxon identified several areas where further company research was needed, including "the impact of anthropogenic sources on climate" and "the potential greenhouse effect."

As with the Committee on Smoke and Fumes' work during an earlier generation's debate over smog in Los Angeles, the point of this science was to prepare Exxon for political warfare. "We should determine how Exxon can best participate in all these [scientific] areas and influence possible legislation on environmental controls," the company wrote. Nearly a decade before climate change gained mainstream awareness in the United States, Exxon was already warning that "it is important to begin to anticipate the strong interventions of environmental groups and be prepared to respond with reliable and credible data."

The threat posed by such groups was real. As Imperial put it in another corporate document from this time period: "The efforts of small groups of hard-core activists have been remarkably successful in postponing and delaying" energy projects.

To that end, Exxon made studying climate science a top priority. It hired scientists and mathematicians to create models predicting future changes to the climate. In 1979, Exxon spent $1 million on a project that transformed a supertanker into a floating research lab. Sailing from the Gulf of Mexico to the Persian Gulf, the tanker took measurements of carbon dioxide in the air and water. The goal was to determine how much greenhouse gases the ocean could absorb. Knowing the answer to this question was important to Exxon, because it would potentially determine at what point the company's fossil fuels would cause irreparable atmospheric damage and force it to change its business model. A senior Exxon scientist would later tell the media outlet *Inside Climate News* that when he arrived at the company in 1981, "I was quite surprised to discover that people in the research lab were very aware... of the greenhouse effect."

Imperial Oil was also aware of this climate research, which was frequently shared between the Canadian subsidiary and its better-known parent company. "It is assumed that the major contributors of CO_2 are the burning of fossil fuels... and oxidation of carbon stored in trees and soil humus," Imperial wrote in a 1980 report that was circulated to Exxon managers and advisers in Houston, New York, and London. "There is no doubt that increases in fossil fuel usage and decreases in forest cover are aggravating the potential problems of CO_2 in the atmosphere." The company could even put precise figures to the problem. It reported that in the year 1850, carbon dioxide concentrations in the atmosphere were 280 parts per million. By 1978, Imperial noted, those concentrations had risen to 330 parts per million.

The company also knew that if carbon emissions kept on rising at the current rates, the consequences could be severe. Just how severe was sketched out during a 1980 meeting of the "CO_2 and Climate Task Force," a research group set up to disseminate climate research within the oil and gas industry. The task force included scientists from Exxon and other U.S. companies, as well as the American Petroleum Institute, which counted among its members Imperial's main oil sands competitor, Suncor. During that meeting, industry scientists predicted that "a 3% annum growth rate of CO_2," meaning carbon emissions grow by 3 percent every year, could cause the average global temperature to rise by 2.5 degrees Celsius (4.5°F). This in turn "brings world economic growth to a halt in about 2025." Participants were well aware that if that cataclysmic outcome took place their products would be to blame. There is "strong empirical evidence," the task force noted, that the warming they were observing was caused "mainly from fossil fuel burning."

Within Exxon there continued to be scientific debates about how worried the company should be about climate change. In the early 1980s, one scientist, named Werner Glass, speculated that future warming wouldn't get as extreme as some of the models predicted because increased carbon would result in more cloud cover and other changes that would slow down temperature rise. Therefore, Glass concluded, the global impacts would be "of a magnitude well short of catastrophic." But Exxon research manager Roger Cohen wasn't so sure. "I think this statement may be too reassuring," he wrote in a company memo to Glass. Cohen agreed with Glass that climate impacts by the year 2030 "are likely to be 'well short of catastrophic.'" However, the world by that point could be hurtling toward a terrifying future—that specific year might simply be the relative calm before a civilization-throttling storm. "It is distinctly possible," Cohen wrote, that warming caused by the industry's products "will later produce effects which will indeed be catastrophic (at least for a substantial fraction of the earth's population)."

Yet even with these increasingly dire warnings, Imperial Oil kept expanding production in the oil sands. From 1983 to 1988, its Syncrude consortium spent $1.6 billion to increase the supply of bitumen from northern Alberta, causing output to reach a new high of fifty million barrels per year. Without an explosion that shut down its main plant for months in 1984, that number likely would have been higher. Meanwhile, the other major oil sands producer was also increasing production. The 1980s were a tough period for Suncor, owing to a major fire at its own plant, the death of its CEO in a plane crash, and the collapse of crude oil prices. Nonetheless, Suncor managed to grow its own production of climate-warming bitumen to a record high of 55,015 barrels per day by the end of the decade.

By the mid-1980s, both these companies were heavily invested in an industry that faced a barrage of threats: from polluted First Nations communities to aggressive activist groups and from swings in public opinion to government policy that would make it costlier to extract oil. In the worst case, Imperial and Suncor could be forced to transition away from fossil fuels entirely. In this battlefield, it was in the best interest of companies to have advance knowledge of emerging environmental issues. The major oil companies could all see that climate change was the next big one. As Ken Croasdale, a Calgary-based scientist who researched global heating for Exxon during this period, explained to the *Los Angeles Times*, the company "should be a little bit ahead of the game trying to figure out what it was all about."

"Very strong interests at stake"

IN LATE SEPTEMBER 1980, viewers who tuned into ABC's *20/20* news program learned about an alarming new threat to public health. During the thirty-minute segment, a youthful-looking Geraldo Rivera, then a prominent investigative journalist, explored the subject of train derailments and hazardous chemical spills. Rivera interviewed railroad employees who said they feared their lives were at stake as they helped move trains carrying dangerous materials along deteriorating rail tracks. "Working on the Railroad," as the ABC report was titled, had all the elements of successful public service journalism: gripping personal stories, attention to an under-the-radar public health issue, a call to action to prevent further disasters.

But employees at the oil, gas, and petrochemical company Shell whose job it was to monitor public opinion on environmental issues recoiled when watching the Rivera segment. "How has this issue impacted the chemical industry? It already has had considerable impact in the form of unfavorable publicity which may result in costly and restrictive regulations regarding the shipment of chemicals," explained a 1980 article referencing Rivera's journalism in a bimonthly Shell publication entitled *TREND*. "Meanwhile, derailments are likely to continue—generating more publicity, followed by Congressional investigations."

Like Charles Koch and many others in the oil and gas industry, executives at Shell fretted about what they saw as an era of government overreach in the economy. They believed the power of companies was being eroded by overzealous regulators and anxious consumers easily convinced by what they watched on TV. "There are times when the public and the government must act as a containing influence on business, but the trend of the last 20 years has been for the government to intrude into the private sector beyond the point where it is proper or beneficial," said Philip Carroll, vice president of Shell's Public Affairs unit, in *TREND*.

Carroll's organization was set up within the company to take a more proactive approach to combating publicity that could hurt Shell's profits. Public Affairs operated on the principle that once an issue reached the attention of Congress, it was already too late. "For example, energy policy is formed as people talk to legislators. And legislators talk to one another. At home. Because we fail to become involved in these early stages, we are forced to react—in Washington—as policy suddenly appears that is not in the nation's best interest as we see it," explained Shell Chemical president James B. Henderson.

To avoid these scenarios, Shell said in the corporate publication, the company would need to be able "to detect issues as they emerge." That's where the Public Affairs unit came in. "By monitoring network television, magazines and newspapers, I think we can anticipate and become proactive to those [issues] which may have major impact on our business in the 80s and 90s," Carroll said. The process generally went as follows: once an emerging issue was identified, a Public Affairs policy development specialist put together a team of experts at the company. That team would "set about developing a Shell position on the issue and plan for action." Crucial to this effort was that Shell understand the issue inside and out: "This usually involves a great deal of research."

It was around this time that Shell began researching the emerging issue of climate change. In 1981, the company commissioned the University of East Anglia in England to undertake "a study of the greenhouse effect." The school, one of the world's leading scientific research institutions, had recently created a Climate Research Unit led by Dr. T. G. Wigley. With an initial grant of £10,000, Wigley's unit began a multiyear investigation into the causes and consequences of global temperature rise. Four years into the project, a Shell representative named M. H. Griffiths paid a visit to see how things were going.

"I found Wigley very much had his feet on the ground," Griffiths reported to Shell about the climate scientist:

> [He] was at great pains to emphasise the uncertainties that still exist in this area and the time needed before which it will be possible to reach any very definite conclusions about the greenhouse gas effect. Having said that, he was prepared to stick his neck out and say that there has been a global warming over the last 100 years, that the 0.5 degrees [Celsius]... increase is a result of

> CO_2 build-up, that we will see a further 1–2 degree warming over the next 40 years and that the warming will be greater in higher latitudes and more in winter than in summer. Such a rise would be greater than any change in the last 1000 years.

If the warming eventually reached four degrees Celsius (7°F), the Shell representative reported, it "might result in the disappearance of all Atlantic sea ice in the summer months" and cause sea level rise of up to four feet. Changes to the climate would likely affect tropical developing countries such as the Philippines, the company learned: "Monsoon rainfall ought to increase also as should the frequency of tropical storms which is again a temperature dependent phenomenon." Putting precise figures on all this was difficult to do, Griffiths reported emphatically, because this is all "a very new science!" But there was one area in which Wigley provided more reassuring news to Shell. In the East Anglia researcher's opinion, the science on climate change wasn't advanced enough yet to compel government action. "He believes that it is not realistic to make policy decisions aimed at reducing the effect of CO_2," Griffiths wrote.

That was good for Shell, because it was at that moment looking to join Suncor, Imperial Oil, Koch Industries, and others as a major player in the Canadian oil sands. By the early 1980s, the European oil giant was leading a $13 billion proposal to extract and refine the hard-to-process oil. The project was called Albian Sands, so named for the Albian Boreal Sea, which had existed 100 million years ago in what is now the Fort McMurray region. But Shell was running into the same problems as Imperial and Sun Oil: cost overruns, technical difficulties, extreme weather conditions, and fluctuations in the global price of oil, all of which threatened to make its massive investments uneconomic.

Shell eventually decided to scale back the oil-producing aspects of its plan for the oil sands and focus instead on building a $1.4 billion refinery outside of Edmonton, Alberta. This project would process fifty thousand barrels of oil per day from the oil sands, turning it into gasoline, diesel, and jet fuel. Though Koch Industries was at this time processing more and more crude from the oil sands, Shell's new refinery was the first in the world designed specifically for the oil sands industry.

The Scotford refinery, a massive complex of pipes and industrial machinery visible from far away on the flat Alberta prairie, opened in September 1984. The "crowded official opening" was covered by the *Calgary Herald* newspaper, which reported that Alberta premier Peter Lougheed's voice was "filled with emotion" as he talked about the jobs and economic prosperity he predicted the refinery would bring. "He praised [the Canada] wing of international Royal Dutch Shell for keeping faith with Alberta by building its biggest-ever project, the largest private investment made in Canada during the recession [of the 1980s]," the *Herald* wrote.

Shell knew full well this project would accelerate the warming of the atmosphere, leading to potentially catastrophic damage across the planet, threatening the lives of millions. This is according to a 1988 report from the company's Greenhouse Effect Working Group synthesizing what it was learning about climate change. In the "confidential" document, which was unearthed by Dutch journalist Jelmer Mommers of *De Correspondent*, the company acknowledged that "Man-made carbon dioxide, released into and accumulated in the atmosphere, is believed to warm the earth through the so-called greenhouse effect. The gas acts like the transparent walls of a greenhouse and traps heat in the atmosphere that would normally be radiated back into space." This could eventually result in warming of up to 3.5 degrees Celsius (6.3°F), Shell noted.

Shell knew that its products were the lead driver of this emissions growth, stating that "the main cause of increasing CO_2 concentrations is considered to be fossil fuel burning. Only fossil fuel burning can be fairly accurately quantified." Analyzing the latest climate data, Shell concluded that global carbon dioxide emissions grew nearly 5,700 percent between 1860 and 1981. "In 1981," it concluded, of the 5.3 gigatons of carbon dioxide the world released,

"44% came from oil, 38% from coal, and 17% from gas." Shell went even further than that, however. Its report attempted to quantify the company's specific contribution to climate change. In the early 1980s, it concluded, Shell was responsible for 4 percent of the world's total emissions from fossil fuel combustion, a massive contribution for a single company.

The planetary changes that Shell was contributing to were unprecedented in human civilization. The company observed that

> Mathematical models of the earth's climate indicate that if this warming occurs then it could create significant changes in sea level, ocean currents, precipitation patterns, regional temperature and weather. These changes could be larger than any that have occurred over the last 12,000 years. Such relatively fast and dramatic changes would impact on the human environment, future living standards and food supplies, and could have major social, economic and political consequences... The changes may be the greatest in recorded history. They could alter the environment in such a way that habitability would become more suitable in the one area and less suitable in the other area.

Shell gave a chilling example. "More than 30% of the world's population live within a 50-kilometre [30-mile] area adjoining oceans and seas, some even below sea level. Large low-lying areas could be inundated (e.g. Bangladesh) and might have to be abandoned or protected effectively," it predicted. At this point, the document shows, Shell was aware that entire countries might have to be someday abandoned due to the burning of its products.

The oil company didn't necessarily think that quick solutions were required. "The likely time scale of possible change does not necessitate immediate remedial action," Shell observed in the 1988 document. "However, by the time the global warming becomes detectable it could be too late to take effective countermeasures to reduce the effects or even to stabilise the situation." This is a point repeated several times in the document. "With the very long time scales involved, it would be tempting for society to wait until then before doing anything. The potential implications for the world are, however, so large that policy options need to be considered much earlier. And the energy industry needs to consider how it should play its part."

Shell was concerned about how the public would react once awareness of the massive consequences of burning fossil fuels became more widespread. Like Exxon, it predicted that environmental concerns about air and water pollution could soon give way to global concerns about the climate. "At present, the focus is on acid rain and nuclear energy," the company noted. "It is possible that perception of a serious environmental threat could swing opinion away from fossil fuel combustion." Shell knew that cutting the amount of oil and gas it produced and refined would go a long way toward addressing the crisis. "An overall reduction in fossil fuel use would of course reduce CO_2 production and could be achieved by constraint on energy consumption, by improved thermal efficiency and by replacing fossil fuels with e.g. nuclear energy," it wrote.

The company didn't think it should have "to take the main burden" because governments and consumers were also contributing to global temperature rise. "But [Shell] has very strong interests at stake and much expertise to contribute," the document stated. "It also has its own reputation to consider, there being much potential for public anxiety and pressure group activity."

This was a period of uncertainty for the company. It now realized that its products were causing atmospheric changes so profound as to "have a substantial impact on global habitability." But the company had just spent $1.4 billion building the world's first oil sands refinery—and it had plans to get much more deeply involved in the industry and expand its production capabilities globally. Shell knew it was only a matter of time before it would be forced to respond to public anxiety and outrage about the environmental consequences of this business strategy. Wait too long, it feared, and it would be forced to react to political forces outside of its control. The 1988 document urged Shell's leaders to start thinking strategically about the political battles ahead: "With fossil fuel combustion being the major source of CO_2 in the atmosphere, a forward looking approach by the energy industry is clearly desirable." Shell needed to be ready for a fight.

"Pitted against our very survival"

JAMES HANSEN was visibly sweating. It was a late June day in 1988 and the NASA scientist was about to begin speaking to a Senate hearing on climate change. Hansen, wiping his brow, had good reason to feel nervous. There were two rows of television-camera lights pointed directly at him, and every available seat in room 366 of the Dirksen Senate Office Building was taken. Hansen had earlier told colleagues that "I'm going to make a pretty strong statement." But the perspiration on his forehead wasn't necessarily due to nerves. Hansen was testifying that day during what was then the hottest and driest summer in global history.

"Everywhere you looked, something was bursting into flames. Two million acres in Alaska incinerated, and dozens of major fires scored the West. Yellowstone National Park lost nearly one million acres. Smoke was visible from Chicago, 1,600 miles away. In Nebraska, suffering its worst drought since the Dust Bowl, there were days when every weather station registered temperatures above 100 degrees," Nathaniel Rich later wrote in the *New York Times Magazine*, recounting the events that led up to Hansen's testimony. "Harvard University, for the first time, closed because of heat.

New York City's streets melted, its mosquito population quadrupled and its murder rate reached a record high."

To many people it must have all seemed like a freak weather event, uncomfortable and extreme but ultimately random. Hansen knew otherwise. The mild-mannered scientist from Iowa had since the 1970s been conducting research in the burgeoning field of climate science. In the same period when Exxon was secretly measuring atmospheric concentrations of carbon dioxide and learning that its products were to blame for global warming, Hansen was reaching similar conclusions. He was among the first scientists to predict that if we kept burning fossil fuels at our current rate, the world could heat up by a civilization-destabilizing 2.5 degrees Celsius (4.5°F) within the coming century.

Hansen had persevered through heavy budget cuts to his research by the Republican administration of Ronald Reagan. And even as he prepared his speech for the 1988 Senate hearing, colleagues at NASA wondered if it was responsible for Hansen to tell the public that global warming was already happening and not just some far-off threat in the future. But Hansen had the support of senators including Timothy Wirth, a Democrat from Colorado, who on the day of the hearing told reporters they should prepare for a major news event. As journalists and observers filed into the room, Wirth urged people to claim seats. "There is no point in standing up through this on a hot day," he said, according to Rich's account of the hearing.

With the crowd seated and eager to hear what the scientist had to say, Hansen began his testimony. Barely raising his eyes as he read from his notes, Hansen argued "with 99 percent confidence" that humankind was already altering the atmosphere through its rising emissions of greenhouse gases. "It is changing our climate now." In a press scrum following his testimony, Hansen was even more direct: "It is time to stop waffling so much and say that the evidence is pretty strong that the greenhouse effect is here."

His remarks made front-page news the next day in the *New York Times*: "Global Warming Has Begun, Expert Tells Senate." Just below that unprecedented headline was another: "Sharp Cut in Burning of Fossil Fuels Is Urged to Battle Shift in Climate." The *Times* story reported that "Until now, scientists have been cautious about attributing rising global temperatures of recent years to the predicted global warming caused by pollutants in the atmosphere, known as the 'greenhouse effect.'" But those days were over. "If Dr. Hansen and other scientists are correct," the *Times* wrote, "then humans, by burning of fossil fuels and other activities, have altered the global climate in a manner that will affect life on earth for centuries to come." Dozens of major newspapers and TV programs carried similar coverage.

Just like Exxon and Shell had feared, some Democratic lawmakers in Congress saw newfound media interest in climate change as an opportunity to bring in regulations on polluting industries. If the public was sufficiently concerned about the issue, then it would be less risky for politicians to take action. "Nobody wants to take on any of the industries that produce the things that we throw up into the atmosphere," Democratic senator Dale Bumpers reportedly explained. "But what you have are all these competing interests pitted against our very survival."

Less than a week after the Senate hearing, hundreds of scientists and policymakers from dozens of countries gathered in Toronto for a conference titled "The Changing Atmosphere: Implications for Global Security." The goal was not only to raise awareness of the climate dangers ahead but to do something about them. A closing statement from the weeklong event explained what was at stake: "Humanity is conducting an unintended, uncontrolled, globally pervasive experiment whose ultimate consequences could be second only to a global nuclear war." Delegates said that to get the emergency under control, countries around the world would need to cut their current carbon dioxide emissions one-fifth from current levels.

By necessity that would mean massive reductions in the production and consumption of fossil fuels.

Within the coming months there were more than two dozen climate change bills working through Congress. Among the strongest was the National Energy Policy Act proposed by Senator Wirth, which if enacted would have forced the United States to cut its energy use by 2 percent every year for the next decade and a half. Meanwhile, mainstream concern about global warming was taking off. The following year, environmental writer Bill McKibben published *The End of Nature*, now considered the first book about climate change written in non-scientific language for a general audience. An accompanying feature in the *New Yorker* framed the problem of greenhouse gas emissions in terms both scientific and philosophical:

> Our comforting sense, then, of the permanence of our natural world—our confidence that it will change gradually and imperceptibly, if at all—is the result of a subtly warped perspective. Changes in our world which can affect us can happen in our lifetime—not just changes like wars but bigger and more sweeping events. Without recognizing it, we have already stepped over the threshold of such a change. I believe that we are at the end of nature.

The book became an international bestseller.

But some oil executives acted like none of this was that big of a deal for their industry—at least in public. In January 1989, just six months after Hansen's seminal testimony, the CEO of Imperial Oil, Arden Haynes, gave a speech in Toronto about major trends affecting the oil industry. He didn't mention climate change once. Instead, he projected confidence about a gusher of new oil supplies flowing from Canada into the U.S. "The United States currently imports about 40 percent of its total oil requirement," Haynes explained to a gathering of political, business, and cultural elites known as the

Empire Club of Canada. "Demand is growing quite strongly—last year it increased by about 2 percent."

Canada, Haynes argued, was in a prime position to meet that oil demand. "On the crude oil side, while production of conventional oil from western Canada is generally flat to declining, new areas are coming into play or at least waiting in the wings," Imperial's CEO said. "Our real crude-oil ace in the hole, however, is the oil sands. I'm sure few Canadians realize just how big our oil sands deposits are... With more than a trillion and a half barrels of oil in place, they constitute one of the three truly enormous deposits of crude oil in the world—the others being in the Middle East and in Venezuela, also in the form of oil sands."

Imperial knew full well the climate damage that would come from burning all that oil, even if its CEO Haynes didn't say so in Toronto. The previous year, a management committee put together a report for leaders of Esso, the Canadian trading name for Exxon and Imperial, summarizing the implications of the Brundtland Report, a 1987 document published by the United Nations that for the first time ever defined the concept of "sustainable development." Within that simple phrase, which had been developed following submissions from scientists, NGO leaders, policymakers, and research institutes from around the world, were potentially huge implications for oil and gas producers.

Summarizing the Brundtland Report for Exxon and Imperial leaders, the management committee explained that "all fossil fuels contribute to complex and inter-related pollution problems including global warming," and that this "makes increasing reliance on fossil fuels problematic." Due to these types of concerns, it went on, "we can anticipate continued and probably growing external pressure around the concept of sustained development and a stronger emphasis on conservation and energy efficiencies." With environmental groups certain to build campaigns around these issues, the management committee recommended that Esso "develop strategies, policy and position papers" to defend the company's business interests.

Despite these internal concerns, Haynes assured the well-heeled crowd in Toronto that the future was bright for producers of the country's difficult-to-extract bitumen. "Canada, of course, is extraordinarily blessed in the energy sphere. In this respect as in many others, we're the envy of most other nations." Haynes had harsh words for anyone who disagreed with his assessment of the country's energy sector. "Frankly, I can think of few issues on which so much misinformation is handed out to the public—sometimes innocently as a result of lack of knowledge, but at other times, I believe, by self-serving opportunists with political or ideological axes to grind."

It was a highly ironic statement for Haynes to make. He was effectively describing his own industry's response to climate change. By the end of the decade, Imperial, Suncor, and Shell would help create one of the biggest and most consequential corporate misinformation campaigns in modern history.

III

Solutions Known and Sabotaged

(1988–2002)

"Threaten the existence"

IN 1991, IMPERIAL OIL figured out a way to stop climate change. In one decisive move, the federal government could bring in a policy that would cause coal power plants to shut down, drastically reduce consumption of oil and gas, and spur massive investments in greener technology. It would shift the entire economy away from polluting industries. Levels of greenhouse gases would plummet. Instead of climate-altering emissions rising steadily over the next fifteen years, which was what Imperial predicted would happen without serious action to address to climate change, emissions would be lower in 2005 than they were in 1990. The company foresaw that this solution would achieve "approximate stabilization" of the carbon dioxide pollution causing the climate emergency, according to a company "discussion paper" from 1991 signed by senior executives and later shared with government policymakers.

As with its earlier research on the causes and consequences of global warming, Imperial Oil wasn't doing this work for altruistic reasons. It could see that momentum was building globally for wide-ranging climate solutions and it was worried about the impact that green policies would have on its growing investments in the Alberta oil sands. But in order to figure out the best way to fight back against environmental legislation, Imperial Oil needed to be able to frame the debate in terms favorable to its own business interests. And the best way to do that was becoming an expert on climate policy before everyone else.

In the months and years following James Hansen's 1988 testimony about global warming to the U.S. Senate, things appeared to be moving rapidly. A newly formed Intergovernmental Panel on Climate Change held its first meeting in Geneva. Countries loosely agreed to work toward the goal of reducing global emissions by 20 percent by the year 2005. Meeting in the Netherlands in 1989, more than sixty nations endorsed the idea of creating a binding international treaty to achieve that goal.

In Canada, where Imperial Oil was based, federal action took the form of a $575 million plan to slow down global temperature rise. This was part of a $3 billion "Green Plan" plan to limit air and water pollution and greenhouse gas emissions announced by prime minister Brian Mulroney's Conservative government. By the standards of federal politics, the plan came together rapidly. The government held "extensive, cross-country consultations with thousands of citizens and stakeholder groups" during the summer of 1990. In those meetings people expressed grave fears about the environment. The number who saw it as a "very serious" issue reached an all-time high of 77 percent, according to Gallup polling. The federal government's plan came out less than six months later.

The Green Plan contained stark language about the growing threat of rising temperatures:

> If we continue pumping out greenhouse gases (carbon dioxide, methane etc.) at our present rate, scientists believe we could cause a warming rate of change faster than any other over the past 10,000 years with potentially dramatic impacts. Over the past 200 years concentrations of greenhouse gases in the upper atmosphere have increased faster than at any previous time. Half the CO_2 added to our atmosphere in human history was emitted in just the past 30 years. These greenhouse gases act as a buffer in our atmosphere, trapping the earth's heat. The rising temperatures could mean flooding on our coasts, heat waves and droughts

on our prairies. They could even threaten the existence of some plants and animals.

This warning wasn't coming from an especially progressive government. Mulroney was widely seen as an ideological ally of Republican U.S. president Ronald Reagan—they famously sang "When Irish Eyes Are Smiling" together at a 1985 summit in Quebec City—as well as the hard-line "free-market" government of U.K. prime minister Margaret Thatcher. But Mulroney's green agenda pointed the way toward a different kind of conservativism that was mindful of science and, in the old-fashioned sense of the word, sought to "conserve" an environment facing threat.

The Mulroney government promised to fund a range of solutions that could "stabilize emissions of carbon dioxide (CO_2) and other greenhouse gases at 1990 levels by the year 2000." It promised to reduce the amount of energy required by equipment used in factories; develop new efficiency standards for refrigerators; create stricter building codes; make vehicles burn less gasoline; develop solar energy and fuel cells; plant 325 million trees; change the federal approval process for major projects so that climate impacts were taken into account; pursue an international treaty to limit global emissions; and study the feasibility of bringing in a tax on the carbon dioxide released by major polluters—such as those in the oil sands.

This was not good news for Imperial Oil, as internal documents produced by the company during this period show. It had by this point operated in the oil sands for decades without any controls on the carbon emissions it released, and even then it had been difficult to make its vast projects in northern Alberta profitable. The extreme cold during winter months, the clear-cutting of boreal forest necessary to create vast strip mines, the tarry bitumen that jammed machinery, the intense refining needed to cook oil sands crude into a smooth-flowing substance—it all added costs. "For example," Imperial Oil CEO Arden Haynes said, "the average cost of producing a barrel of light crude in, say, Saudi Arabia, is less than $US 5. The cost of producing a barrel of light synthetic [oil sands] crude at Syncrude is around $US 15."

Those higher costs were caused by higher energy usage. And that in turn resulted in higher greenhouse gas emissions. "To illustrate," Imperial explained in a 1991 report signed off on by CEO Haynes, "the production process for conventional crude oil consumes the equivalent of about 3.5 percent of the energy content of crude oil. This increases to 18.5 percent for crude bitumen." This made Imperial and every other oil sands producer extremely vulnerable to aggressive climate policy. A graph from a 1991 report produced by Imperial showed that the company's annual emissions had spiked from under three thousand kilotons per year in 1984 to nearly seven thousand kilotons by 1988. This, the company said, reflects "a significant volume increase in the production of crude bitumen."

Imperial expected this trend to continue. "Since [the company] produces close to 70% of Alberta's total crude bitumen, a very energy intensive sector, and since the energy requirements in the conventional oil and sectors are increasing due to more production from depleting reservoirs . . . it is expected that the energy requirements, and hence, CO_2 emissions will continue to rise," Imperial explained in a separate document.

Reducing emissions from the oil sands wouldn't be easy, however. Imperial had the data to prove it. Starting in 1991, it ran complex scenarios analyzing the impacts of various climate solutions being discussed at the time. Making large investments in improving the energy efficiency of its operations was a dead end as far as the climate was concerned. It would mostly result in small, incremental carbon reductions that could easily be wiped out if the overall operations expanded. That approach, Imperial concluded, would be "unlikely to stabilize projected CO_2 emissions at 1990 levels."

The same went for capturing the carbon released from its bitumen plants and burying it underground. Imperial calculated that "it would be technically feasible to develop, over a five to 10-year period, the infrastructure to permanently dispose of up to 50,000 tonnes per day of CO_2." But the estimated price tag of "approximately $7.5 billion" for that was way too exorbitant to make economic sense.

Imperial Oil concluded there was one solution that could actually be effective in addressing climate change: a national price on carbon emissions. This aligned the company with the Mulroney government, which in its Green Plan proposed using "taxes and an emissions trading system to attain reductions in greenhouse gas emissions." It commissioned a leading economics consulting firm known as DRI/McGraw-Hill to calculate what would happen if the federal government forced polluters to pay a price of $55 for each ton of carbon dioxide they emitted into the atmosphere. Such a policy, Imperial learned, would go a long way toward stopping climate change. It would result in provinces closing down oil refineries and coal mines. Utilities would build more nuclear plants and expand hydropower facilities. Railroad companies would phase out diesel locomotives. Car engines would be converted to burn natural gas. Cities would build fewer factories and office buildings. Governments would have hundreds of billions of dollars in carbon-tax revenue to spend on conservation programs and clean-energy research.

Instead of rising sharply, Imperial learned, Canada's emissions would plateau around 1990 levels and then begin to shrink. "The results show that of all the policy measures considered, only the carbon tax of $200 per tonne of carbon or $55 per tonne of CO_2 achieves approximate stabilization of Canada's CO_2 emissions." This is exactly the type of action that would have allowed a major world economy to stop climate change.

Imperial calculated this would come with costs. Canada's real gross domestic product, a measure of the scale of economic output, could decline slightly under a moderate-carbon-tax scenario. But the economic hit wouldn't necessarily be felt equally across the entire country. Though provinces like Quebec wouldn't be hugely affected, due to their high reliance on hydroelectricity, places that were reliant on fossil fuels could be hammered. "The Canadian oil and gas industry, which is heavily concentrated in Alberta, would be harshly penalized," the Imperial report concluded.

The oil sands were especially vulnerable. A tax on those emissions could increase "bitumen production costs by about $5 [per barrel]." Because of those higher costs, "heavy oil would virtually cease to be a usable resource," the company warned. That could potentially cost the Exxon subsidiary dearly: "Translating the... carbon tax impacts to Imperial, for example, might result in a 12% reduction in downstream revenue, equivalent to 940M$." The takeaway was obvious: Canada could slow down climate change, or it could tap one of the world's largest oil reserves—but it couldn't do both.

Armed with this knowledge, Imperial began an effort to make the carbon tax look economically reckless. Its targets were government policymakers who had the ability to turn such a tax into official fiscal policy. Imperial's strategy was laid out in a 1993 document labeled "proprietary" that was also intended as a guide for Exxon executives in the United States. The document contained talking points for speaking with "government, thought leaders and media." Imperial would stress the "many uncertainties" associated with taking action on climate change through a carbon tax, arguing the crisis "doesn't warrant drastic steps at this time." It would argue it "makes little sense to act unilaterally to respond to a global issue." Imperial would also attempt to steer the conversation away from the higher-than-average carbon content of its oil sands operations. "Focus should shift to all greenhouse gases, sources and sinks, not just CO_2 production from energy use," it noted.

The document also contained a "Basic Strategy/Action Plan" that could inform Imperial and Exxon's defense against government climate policies. "Stress relative certainty of the debits to Canada's precarious economy and international competitiveness versus the uncertainty in environmental benefits," it read. This was a misrepresentation of Imperial's own research. Though real GDP could decline in a moderate-carbon-tax scenario by 1.8 percent in 1995, its report had found, that might improve to a decline of 0.3 percent by 2005. The reason for that is "government would have enormous amounts of additional revenue once carbon taxes are imposed," which policymakers could use to fund a massive green infrastructure buildout. "The surge in capital spending mitigates the impact on the economy after the year 2000," Imperial acknowledged. And the environmental benefits of taxing carbon were far from uncertain—the company knew that this would be a substantial step toward "stabilizing" climate change.

Thus, at this key moment in the early 1990s when government action could have made a big difference against a mounting global climate emergency, Imperial used its influence to undermine what it knew was the most effective solution. Stopping climate change was not only possible but economically feasible, company leaders knew. But since taxing carbon would kill its profits in the oil sands, Imperial would make sure that this climate solution never happened.

"I feel embarrassed"

ON JUNE 5, 1991, an organization closely connected to Koch Industries held one of the world's first events devoted to climate change denial. Meeting at the Capital Hilton hotel in downtown Washington, D.C., dozens of academics, many of them male, white, and balding, gathered to attack the growing scientific consensus on global warming that James Hansen had helped bring to the public's attention only three years earlier. "The notion that global warming is a fact and will be catastrophic is drilled into people to the point where it seems surprising that anyone would question it, and yet, underlying it is very little evidence at all," claimed one participant. "Nonetheless, there are statements made of such overt unrealism that I feel embarrassed."

"Global Environmental Crises: Science or Politics," as the conference was titled, was hosted by the Cato Institute think tank. The issue of climate change was so new at this point that most conservative organizations didn't have a clearly defined position on it. But the Cato Institute entered the arena of public debate ready to brawl. "Many may find it surprising that respected scientists challenge all of the media-hyped environmental ills," reads a conference brochure. "Concern about global warming, for example, has become a part of the American psyche—despite scientific uncertainty." Attendees saw this as an ominous trend: "We are seemingly marching toward a new world order of population control, economic planning and 'sustainable development.'"

This conspiratorial attack against what was then considered neutral mainstream science was entirely consistent with the Cato Institute's broader mission and strategy—and with the business interests of Koch Industries. When Charles Koch founded the think tank in the mid-1970s, he proceeded to give it as much as $20 million in start-up funding from his own private fortune during its early years. He saw the organization as the generator of ideas that could build support for a radically limited public sector. Government's only legitimate role, Charles said in a speech around this time, is to "serve as a night watchman, to protect individuals and property from outside threat, including fraud. That is the maximum."

The Cato Institute's goal of "lower taxes, looser regulations and fewer government programs for the poor and middle class all corresponded to the Kochs' accumulation of wealth and power," as Jane Mayer writes in *Dark Money*. By the 1990s, Koch Industries had grown into a fossil fuel behemoth, employing thousands of people to run a vast network of oil pipelines across the United States and Canada, in addition to gas plants, petrochemical facilities, coal mines, and automobile dealerships. The "cash cow" of this empire continued to be the company's Pine Bend Refinery in Minnesota, which processed crude oil from Alberta. A Goldman Sachs analysis from the 1980s estimated the value of Koch Industries' refining assets was as high as $1 billion.

As the processing capabilities of Pine Bend grew, Koch Industries was looking to get more actively involved in the Canadian oil sands, where it had held large exploratory mineral leases for decades. The company was considering building an open-pit bitumen mine in the Fort Hills area of northern Alberta, not far from where Imperial Oil and Suncor had their base of operations. The idea was that this new source of heavy oil sands crude could be shipped via pipeline to Pine Bend. "As the demand for heavy crude continues to grow at Koch's Minnesota refinery, we intend to meet that demand with heavy crude from Canada," Koch Exploration president Steve Kromer explained.

Koch Industries, like the other oil companies expanding their operations in the Canadian oil sands, had a keen sense of emerging environmental issues that posed a threat to its profits. In the newfound conversations about climate change happening during the late 1980s and early 1990s, it saw the potential for a vast expansion of government powers culminating in regulations and taxes. "The economic, societal and personal consequences of such changes have been ignored," the Cato Institute's conference brochure argued. "Instead of accepting more regulation and possibly jeopardizing our common future, Cato is bringing together those scientists and policy analysts who offer alternative perspectives on environmental crises."

Opening the conference in a program entitled "Global Warming: Catastrophic? Manageable Change? Beneficial?" was a biological sciences professor from the University of Virginia named Patrick Michaels. The title of his speech, "The Political Science of Global Warming," proved apt. Michaels was not a climate scientist, but he had just spent the previous months working on an experimental media campaign to convince average Americans in three test communities that global warming wasn't real.

That campaign was organized by Southern Energy, a major coal-burning electric utility, as well as the Edison Electric Institute, a utility trade group whose members also relied heavily on coal. These organizations saw in the public's newfound awareness of climate change a threat to their polluting business models. Teaming up with Simmons, a public relations and marketing firm, they conducted public opinion research in Fargo, North Dakota; Flagstaff, Arizona; and Bowling Green, Kentucky. Of five hundred adults interviewed over the phone by researchers, 82 percent claimed to have some knowledge of global warming, a topic that was still quite novel in U.S. media. Worryingly for the coal utilities, more than one-third of respondents backed "federal legislation without any qualification or cost," according to campaign documents that were later leaked.

The goal of the utility-led campaign, which was given the name "Informed Citizens for the Environment," was to shift what people thought and believed about climate change. "Reposition global warming as theory (not fact)," read a strategy document. "Target print and radio media for maximum effectiveness." The optics of coal interests leading an advertising effort to confuse the public weren't great, so they decided to "use a spokesman from the scientific community" to deliver their message. Throughout the spring of 1991, Michaels was their man. The University of Virginia professor met with editors and writers at media outlets across the three test communities and appeared on local radio shows. "We believe it is wrong to predict that higher levels of carbon dioxide will bring a catastrophic global warming," he claimed.

The goal was to make it seem as though these ideas were being generated organically by ordinary people. "Change often begins with one person," Michaels would say after disputing the validity of climate science. "You can make a difference by sharing what you've learned with others." The campaign also hired Rush Limbaugh, the prominent far-right talk radio host, to tape ads that would air with his show. "I know you've been seeing more and more stories about the global warming theory," Limbaugh said in one ad. "Stories that paint a horrible picture. Stories that say the polar ice caps will melt... Well get real! Stop panicking! I'm here to tell you that the facts simply don't jibe with the theory that catastrophic global warming is taking place."

That was the message Michaels delivered to the Koch-backed Cato Institute conference, which took place only a month after his work on the Informed Citizens for the Environment campaign. It was still early days in the climate debate, but Koch Industries had already settled on an effective strategy. Getting a credentialed scientist to poke holes in the rapidly strengthening scientific consensus was a useful way to distract people's attention from the real interests that were threatened by governments taking action on the crisis. "We can look down the road a little way and see an industry under siege," one member of the Koch network, an oil executive from Oklahoma named Lew Ward, predicted during this period. "We are not going to let that happen."

"We have to get this right"

BY 1996, A SMALL GROUP of policy experts became convinced that they had figured out a way to get the climate emergency under control. Though they were little known figures to the broader public, Dan Dudek, Joe Goffman, Fred Krupp, and others associated with a New York City–based organization known as the Environmental Defense Fund, or EDF, nonetheless had an impressive pedigree. In the 1980s, they'd helped convince the George H. W. Bush administration to enact rules limiting the pollution that causes acid rain. It was among the most effective environmental legislation ever adopted in U.S. history. They figured that policy could be proof of concept for a much bolder plan: to make it so expensive and uncompetitive to burn fossil fuels that companies would throw their financial might behind developing less-polluting alternatives.

EDF's plan for stopping climate change had two key components: first, set a hard limit on the level of greenhouse gas emissions that polluters could release into the atmosphere, which they called a "cap." Then, give polluters several options: they could drastically cut their own carbon pollution, or, failing that, they could purchase or receive pollution permits, a new commodity that polluters and nonpolluters alike could "trade." This system of "cap-and-trade" was undeniably complicated. "People would later compare it to a driver making three right turns instead of one left," Eric Pooley recounts in *The Climate War: True Believers, Power Brokers, and the Fight to Save the Earth*. But it had advantages over the types of carbon taxes that Imperial Oil was studying.

"The cap changed behavior," Pooley explains. "Hit a CEO with a new tax and he would bring in his accountants. Hit him with a cap and he would unleash his engineers." This solution had political advantages as well, the experts believed. It didn't carry all the negative baggage of a new "tax" and it left decision-making power in the hands of companies. At the end of the day, it was up to executives to devise ways of eliminating climate-warming gases from their operations or creating brand-new industries that didn't release those gases at all. This made cap-and-trade potentially attractive to politicians across the ideological spectrum.

EDF's experts could point to their past success as a compelling case study. During his 1988 presidential campaign, Bush had promised to take action on acid rain, then a high-profile environmental problem. The crisis was being caused by sulfur dioxide emissions released from the smokestacks of coal-fired power plants across the Midwest. Those emissions reacted with air and water in the atmosphere, creating acidic airborne chemicals that were carried by the wind into Canada and the American Northeast, killing off forests, streams, and lakes. "The time for study has ended," Bush said at a campaign stop at Lake Erie. He promised his Republican administration would cut "millions of tons" of sulfur dioxide.

Once Bush was in office, his advisers requested a meeting with EDF's Fred Krupp, who along with his colleagues Dudek and Goffman had earlier released an influential report called *Project 88*, which argued that environmental problems could be solved by using the creativity and problem-solving skills of the market. "The president is serious about acid rain," one of Bush's advisers reportedly told Krupp. "If you can come up with a market-based approach, we will work with you." There were debates within EDF about whether it was wise for an environmental group to partner with Republicans. But Krupp, according to Pooley's book, was insistent. "It's put-up-or-shut-up time," he said. "We've finally got the chance to go beyond the eight-hundred-word op-ed and figure out how to do this."

Krupp, Dudek, and others brainstormed their plan by scribbling on big sheets of white paper at EDF's headquarters on Park Avenue in New York. They came up with a new paradigm for cutting air pollution. The federal government would set an overall reductions target, and polluters were left to figure out the most cost-effective way of meeting it. The real innovation of the plan was creating permits that polluters could buy if they weren't able to achieve the target, effectively putting a penalty on sulfur dioxide emissions. Egregious polluters would pay a de facto tax, while greener and more innovative companies would gain a market advantage through lower costs than their competitors. President Bush's advisers apparently loved it.

The bill that turned this system into law was debated by the U.S. Senate for ten weeks. During that time, coal-burning electric utilities did their best to kill it, arguing it would cost the industry up to $7 billion per year. But the Bush White House largely ignored them and pressured Congress to follow through. In the months and years following passage of the acid rain plan, sulfur dioxide emissions plummeted as utilities shifted to lower-sulfur coal sources and created better pollution scrubbers for their smokestacks. The costs were far less than the industry had warned, coming in at around $3 billion per year. But there were massive benefits that more than offset those costs. When you factored in the improvements to public health and local ecosystems of keeping toxic chemicals out of communities, lakes, forests, and people's lungs, the acid rain legislation led to an estimated $119 billion in net benefits annually.

All the while, the EDF analysts saw this as a precursor to dealing with global warming. "We have to get this right, because the next major endeavor is climate," Dudek said. "We'll never be able to convince anybody that tradable permits are the best way to deal with [carbon dioxide] unless our real-world experience really works." By the mid-1990s, the sulfur dioxide legislation had been so successful that acid rain was largely fading from the public's consciousness. Instead, as Dudek and others had anticipated, momentum to address climate change was building.

The occasion for scaling up EDF's proposed solution for toxic coal-plant emissions into a plan for stripping carbon out of the economy was a major climate conference set to take place in Kyoto, Japan, in November 1997. For years, delegates from the world's nations had been gathering to figure out an international framework for stopping global warming. An earlier "Earth Summit" in Rio de Janeiro had produced an agreement urging world leaders to stabilize their emissions. But it contained only rough guidelines for how to do so and no firm timetable other than a promise to begin the process "as soon as possible." A subsequent conference in Berlin helped established the principle that wealthier countries should meet mandatory climate goals while giving more flexibility to developing countries like China and India, which had fewer resources and contributed far less to global emissions.

All of this international diplomacy was supposed to produce firm goals, deadlines, frameworks, and penalties at the upcoming Kyoto meeting. As the administration of U.S. president Bill Clinton tried to figure out what its approach would be at the conference, EDF experts began meeting with vice president Al Gore. They made the case that the United States should use the negotiations to push for a global-scale version of the acid rain legislation, but this time targeting carbon. Gore and the chief U.S. climate negotiator, Stuart Eizenstat, were apparently receptive. In talks with European negotiators leading up to Kyoto, Eizenstat insisted that any agreement had to include carbon trading. "Acid rain gave him a brace to lean against," Pooley recounts. "He wasn't digging in on behalf of a theory; he was digging in because cap and trade was already a part of the U.S. system, and it was working."

"Americans can't hear the whistle"

IN JULY 1997, just months before negotiators from 192 nations were set to gather in Kyoto, President Clinton arranged a meeting with six of the country's top climate scientists at the White House. The scientists were brought in to explain the mounting threats posed by global warming. Clinton was joined by vice president Al Gore, thirty members of his administration, seventy-five representatives from industry and NGOs, and the White House press corps. Also present that day was Lenny Bernstein, a chemical engineer for Mobil, an oil company that would merge with Exxon the following year. During the seventy-five-minute event, he took careful notes. Bernstein was there collecting intelligence for the Global Climate Coalition, a powerful group of polluters that was at that same moment trying to convince the public that climate change wasn't real.

Bernstein listened attentively as Oregon State University scientist Jane Lubchenco explained how global temperature rise could make many areas of the United States unrecognizable. Lubchenco described "a New England in which there were no sugar maples, Louisiana with flooded salt marshes, a Midwest which had higher water demand and greater need for pesticides for agriculture and Glacier National Park with no glaciers," according to Bernstein's notes. "She ended with a plea as a Mother to preserve the ecosystems for future generations." The Mobil scientist was apparently unmoved by her presentation. He referred to Lubchenco as "emotional" and bemoaned the fact that "there was no discussion of scientific views which did not support the Administration's position or of the cost of responding to climate change."

As Bernstein scribbled away, President Clinton explained that the time for questioning whether global temperature rise was happening was over. "He characterized climate change as no longer a theory, but 'for real', and [said] that there was ample evidence that human activities are already disrupting climate," Bernstein wrote. But the president acknowledged there was more to be done to convince ordinary people to take the crisis seriously. "He said that 'we (the Administration) can see the train coming but most Americans can't hear the whistle blowing,'" the Mobil engineer wrote. Clinton closed the meeting with a call to action, "saying that this was the beginning of a consistent, long-term effort to involve the people of the country in the issue." The president argued that "the quicker we start, the less extreme the solutions will be."

Bernstein had spent years immersing himself in the latest science on climate change, and he knew that what Clinton was saying about the crisis was true. But four months later, a series of ads paid for by Bernstein's organization, the Global Climate Coalition, began trashing just about everything Clinton had said during the meeting. "On December 1st, there will be a United Nations meeting in Kyoto, Japan. The U.S. will be asked to sign a global climate treaty that could increase our energy costs by 20% or more," read a full-page spread in the *New York Times*. "That won't help the environment, but it will hurt America's economy." The ads directed readers to visit a website called "Climate Facts." Once there, readers would learn that "there is conflicting data over whether there is a warming trend." Another section, entitled "Are humans causing the warming?," claimed that "the sun may be playing a larger role in the Earth's climate than was previously suspected."

The Global Climate Coalition was formed in 1989, just one year after James Hansen's testimony to the Senate about climate change. Its early membership was a cross-section of America's polluting corporate elite. It included car manufacturers (General Motors, Chrysler, and Ford); coal miners (BHP-Utah International and Peabody); fossil-fuel-burning utility companies (Edison Electric); and business trade groups (the Business Roundtable and the U.S. Chamber of Commerce). It also included oil and gas producers with a direct financial stake in the Canadian oil sands. One of the Global Climate Coalition's founding members, Shell Oil, had opened its $1.4 billion bitumen refinery in Alberta only five years earlier. Suncor, Imperial Oil, and other oil sands heavyweights were represented in the coalition through the American Petroleum Institute, another founding member.

"I would like to introduce you to the Global Climate Coalition," the organization wrote to U.S. Republican senator John Heinz several months after its founding. "We are an association of businesses and business organizations that are working with scientists, policymakers and others to responsibly address global climate issues." The letter told Heinz to expect a visit from a coalition member within the next several months. During the following years the Global Climate Coalition facilitated meetings with dozens of U.S. senators and other policymakers.

An early bulletin produced by the coalition may give some sense of what was discussed in those meetings. "[Scientists'] predictions of a climate apocalypse were based solely on computer models," it reads. "These simulations were used by many as a rationale to call for immediate government action to restrict the burning of fossil fuels. Today, just a few years later, we are nearing the end of one of the coolest years on record in the eastern United States. Many scientists have growing doubts about the accuracy of those forecasts that captured widespread media attention and made global climate change a significant international policy issue."

By the mid-1990s, the Global Climate Coalition was regularly getting academics who questioned the scientific consensus on climate change quoted in major media outlets. Those contrarians included Patrick Michaels, the University of Virginia professor who spoke at the early Koch Industries–backed climate denial conference in Washington, D.C. A *Washington Post* story from 1993 referencing Michaels's work proclaimed that the "greenhouse effect seems benign so far," explaining to readers that "in summer, when stress is hardest on living things and when ice caps melt, temperatures are no warmer than they were in the 1860s and 1870s." The Global Climate Coalition was thrilled. "*Post* reporter finds threat uncertain," it reported in its newsletter.

Privately, the coalition knew that Michaels and the other global warming contrarians it helped to get quoted were not producing credible science. Bernstein, the Mobil engineer who sat in on President Clinton's climate change meeting, acknowledged as much in an internal scientific primer for Global Climate Coalition members that he authored in 1995. "Prof. Patrick Michaels of the University of Virginia presented a series of hypotheses about how greenhouse warming should affect temperature," Bernstein wrote. "The contrarian theories raise interesting questions about our total understanding of climate processes, but they do not offer convincing arguments against the conventional model of greenhouse gas emissions–induced climate change."

Instead, Bernstein wrote, "the scientific basis for the Greenhouse Effect and the potential impact of human emissions of greenhouse gases such as CO_2 is well established and cannot be denied." Yet that was not the message the organization was presenting to the public. The same month that Bernstein's primer was making the rounds internally, the coalition's "Year-End Climate Watch Bulletin," a publicly circulated newsletter, informed readers that "warming that has occurred to date... is within the range of natural variability... no credible scientific evidence exists which shows that these changes have been caused by human activity."

The Global Climate Coalition's disinformation campaign came during a tenuous moment for global efforts to stop climate change. In the lead-up to Kyoto, there was international momentum building for an agreement that could compel countries to stop their greenhouse gas emissions from growing and shrink those emissions below 1990 levels. And across the world there were many creative solutions being discussed for how to achieve this without hurting the economy—the acid-rain-inspired "cap-and-trade" framework that policy experts from the Environmental Defense Fund had discussed with Vice President Gore was just one example.

But the Berlin conference in 1995 revealed fault lines in terms of which countries would lead on reducing emissions. Clinton had agreed during those talks to give China, India, and other developing countries more leeway as a matter of fairness. Countries like the United States had gotten rich from burning fossil fuels and caused the majority of atmospheric warming to date, so they had the resources and moral imperative to move first. Back in the U.S., however, Democratic senator Robert Byrd, from the coal state of West Virginia, proposed a resolution in July 1997 claiming that giving any special treatment to China at the Kyoto talks would "result in serious harm to the economy of the United States." He was backed by senator Chuck Hagel, a Republican from Nebraska and close ally of the Global Climate Coalition, who in his floor speech referenced the dubious science of Patrick Michaels as evidence that climate fears were overblown. The resolution passed unanimously: Congress wouldn't ratify any climate agreement that gave developing countries laxer goals and timelines than the United States.

"Byrd-Hagel was the Senate's first chance to speak on global warming, and it spoke with a loud, unambiguous *hell no*—framing the issue as a choice between the earth and the economy, and making clear that economy came first," Eric Pooley recounts in *The*

Climate War. "The notion that fixing the climate necessarily means destroying the economy was to become the Big Lie of the climate debate and the signature achievement of the opponents of action."

There is evidence that this lie partially originated in Canada. When climate change first started becoming a major public issue, in the late 1980s and early 1990s, it wasn't automatically assumed by politicians, even conservative ones, that stopping the warming of the atmosphere would destroy the economy. Republican president George H. W. Bush had shown that fixing the problem of acid rain could bring an estimated $119 billion in benefits to the U.S. economy, and in Canada the Conservative prime minister Brian Mulroney had played a lead role in convening the Montreal Protocol, an international agreement that successfully brought under control the emissions responsible for destroying the ozone layer. A later study by Environment Canada found the Montreal Protocol's "economic benefits exceed costs by some $224 billion."

But in the early days of mainstream climate awareness, there were relatively few people researching which policies could be most economically effective at fixing the crisis. Because they had studied climate change internally for decades, oil companies were in a prime position to provide credible-looking research on solutions, which then allowed them to define the debate in terms favorable to their business model. This is precisely what had happened when the oil sands producer and Exxon subsidiary Imperial Oil commissioned one of the first-ever studies calculating the impacts of a carbon tax. That study predicted "the overall cost to the Canadian economy of such a carbon tax would be a cumulative reduction of about $100 billion (real) in the gross domestic product."

These dire warnings mostly focused on the heavy financial hit to oil production from carbon regulations and largely ignored the potentially massive economic benefits that would come from governments spending big on green infrastructure (not to mention all the money saved from avoided deaths and illness caused by the oil and gas industry's air and water pollution). Nonetheless, Imperial's early carbon-tax study was specifically referenced in Canada's 1994 National Report on Climate Change, a document submitted to the United Nations.

The consulting firm behind that Imperial study, DRI/McGraw-Hill, would several years later run a similar analysis for the Global Climate Coalition. "U.S. living standards and lifestyles would be seriously damaged by many of the greenhouse gas abatement proposals currently under consideration, especially those that would stabilize or reduce carbon emissions by taxing fossil fuels," reads a 1996 background paper released by the coalition. That paper cited a "DRI/McGraw-Hill study of carbon taxation," which found that such a policy could destroy one million jobs within the tax's first two years of implementation. The use of the same consulting group wasn't the only link to the Canadian oil sands. At that point Imperial Oil's parent company, Exxon, was a lead coordinator of the Global Climate Coalition.

These connections meant that ideas innovated in Alberta's oil patch were amplified south of the border. "Imperial warned Canada not to go it alone in cutting emissions, emphasizing that Canadian policies would not solve a global problem and would hurt the Canadian economy," the Climate Investigations Center observed of the Imperial carbon-tax study from 1991. "This rhetorical tactic was used by the Global Climate Coalition circa 1997 to apply pressure in the U.S."

In the weeks leading up to Kyoto, the coalition blitzed mainstream media outlets across America with warnings about the grave economic dangers of reducing emissions. "The United Nations is negotiating a climate treaty that will require severe restrictions on the amount of energy we use," another full-page ad in the *New York Times* claimed. "And it puts the entire burden on the U.S. and a few other countries." Meanwhile, a TV ad broadcast during this time showed a middle-aged man filling up at the gas station. Over music that could have been lifted from a mid-1990s psychological thriller, a female narrator claimed that acting on climate change would "force the U.S. to cut energy use by over 20 percent, gasoline prices could go up by 50 cents a gallon, heating and electricity prices could soar."

The coalition and its member companies continued to hammer the scientific consensus as well. "Let's face it: the science of climate change is too uncertain to mandate a plan of action that could plunge economies into turmoil," read 1997 ads taken out by Mobil in the *Washington Post* and the *New York Times*, right around the time that Mobil's engineer Bernstein was attending the White House climate meeting convened by Bill Clinton. "Scientists cannot predict with certainty if temperatures will increase, by how much and where changes will occur."

To the average person this was all deeply confusing. On the one hand they were hearing from climate scientists and United Nations representatives that if we didn't take serious action to limit greenhouse gas emissions, the result could be unimaginable global pain and suffering—and on the other they were reading advertisements saying the science was unsettled and any government action would hurt them personally. The problem wasn't that people couldn't hear the whistle of an approaching crisis; it was that they were being blasted with distracting noises from every possible direction.

By some measures the Global Climate Coalition's disinformation campaign of the 1990s was hugely influential. A major study of global opinion on climate change later conducted by several U.K. universities found that in 1989, nearly 65 percent of Americans worried about climate change a "great deal" or "fair amount," But by 1997, that number had fallen to 50 percent. "Through the 1990s, at a critical point when the fossil fuel usage needed to be brought under control, public concern about the risks and causes of climate change waned," reads a separate academic study summarizing public opinion research from that time period.

During eleven days of intense negotiations in Kyoto in late 1997, Vice President Gore and the U.S. negotiating team did their best to help create an agreement, even as public opinion back home about the need for action was faltering. In addition, Pooley writes in *The*

Climate War, Gore and his team "were acutely aware that they were striking a deal that the U.S. Senate would never ratify," due to the resolution supported by the Global Climate Coalition that had been proposed by Senators Byrd and Hagel earlier that summer. The final Kyoto agreement contained a fledgling promise to work toward a global system of cap-and-trade for reducing greenhouse gases. But that was severely undermined by Clinton's failure to convince the Republican-led Senate to ratify the Kyoto Protocol.

When George W. Bush won the 2000 U.S. election, he had no interest in trying. "As you know, I oppose the Kyoto Protocol because it exempts 80 percent of the world, including major population centers such as China and India, from compliance, and would cause serious harm to the U.S. economy," President Bush explained in a 2001 letter sent to four Republican senators.

The similarities between this statement and the disinformation that the Global Climate Coalition had been amplifying for more than a decade was no coincidence. A State Department memo from 2001 containing talking points that a government official should use in an upcoming meeting with the coalition was explicit: "POTUS [aka President Bush] rejected Kyoto, in part, based on input from you."

"The dumbest-assed thing"

IN NOVEMBER 2002, Canadian journalists received an invitation to an unusual press event that was taking place in Ottawa. "You are invited to attend a News Conference where, for the first time in Canada, a group of the world's leaders in climate science and energy engineering will be assembled to reveal the science and technology flaws of the Kyoto Accord," it read. Journalists who showed up would hear twenty-five academics with impressive-sounding credentials attack the climate agreement that Canada had joined five years earlier in Japan. Among the day's speakers was a veteran spreader of climate disinformation: Patrick Michaels.

For more than a decade Michaels had been fine-tuning his pitch about why addressing climate change was a terrible idea. He now had it down to a science. "Global warming is overblown," he'd recently claimed in an article trashing the Kyoto Protocol that was published by the Virginia Institute for Public Policy, a free-market organization with financial links to Koch Industries. "Historical records show that about two-thirds of this warming will be in the cold portion of the year, and the lion's share will be in the coldest, most deadly air," he wrote. "Does this sound like something we should spend a fortune trying to stop?"

Michaels was joined at the Ottawa event by more than two dozen others outspoken in their belief that Kyoto was a financially disastrous way to address a climate crisis that didn't even exist. The event was aimed at policymakers as much as the general public. Canadian Members of Parliament were set to vote the following month on whether Canada should officially endorse a protocol calling for the country to cut its greenhouse gas emissions by 20 to 30 percent within a decade. "There's a lot of people who are prepared to vote on this even though they don't have enough information, and that's not a very responsible thing to do," claimed Evan Zelikovitz, whose public affairs firm APCO Worldwide organized the "Kyoto's Fatal Flaws Revealed" event.

But the man who bore the most responsibility for hosting a parade of U.S. climate change deniers in the Canadian capital was Robert Peterson, chairman and CEO of Imperial Oil. Contrarians like Michaels were ultimately a noisy distraction, and so were the APCO employees who handled the event's logistics and got quoted in the media. The corporate sponsorship that made the event possible was provided by Imperial. And it was no mystery why Imperial wanted the public to be confused about climate change. Confusion meant uncertainty, and uncertainty meant paralysis. That in turn meant Imperial could keep pulling climate-destroying bitumen out of the earth without having to worry about expensive carbon regulations.

The Kyoto Protocol, Peterson insisted during the lead-up to the Ottawa event, had "been couched and clothed in some kind of environmental movement." What Peterson saw was a direct attack on the polluting fuel that he'd devoted his life to extracting and selling. He'd use any dirty trick available to fight back.

That was not how Peterson presented himself to friends, family, and colleagues, however. They saw him as a person of strong morals, a born leader who "delivered a constant message of leadership, ethics, governance and social responsibility," someone "known for his ambition fed by his enduring preparation and thoughtfulness." That's according to Peterson's 2021 obituary, which noted that the oil sands executive "derived joy in emphasizing education" for his grandchildren. It didn't note that Peterson was also a spreader of dangerous lies about the global temperature rise caused by his company's business model.

Peterson was born on the prairies in Regina, Saskatchewan. He got a job as a summer student at Imperial in 1958 and never left, eventually rising to the position of chief reservoir engineer. During the uncertain years when Imperial was developing its first oil sands projects, Peterson was steadfastly in favor. "He championed the company's 1970s leadership in the Alberta oilsands both as the senior partner in the Syncrude Canada consortium and with its own Cold Lake project," reads an entry on Peterson in the Canadian Petroleum Hall of Fame, an honor bestowed on those who bring "success and prosperity" to the industry.

Peterson became chairman and CEO of Imperial in 1992, just as the global movement to prevent climate breakdown was gaining momentum. Meeting in Rio de Janeiro and New York that year, 165 countries had agreed to create the United Nations Framework Convention on Climate Change, "acknowledging that change in the Earth's climate and its adverse effects are a common concern of humankind." Peterson was suspicious from the start. "The [U.N.] framework has turned out to be as much, or more, about world trade, future competitiveness and economic growth than issues of potential global climate change," Peterson wrote in 1995 in a private letter to then Canadian prime minister Jean Chrétien.

Several years later, Peterson was openly attacking the science behind climate change. In a 1997 letter to shareholders, Imperial raised alarms about the "limiting or restricting [of] fossil-fuel consumption" that could result from the upcoming Kyoto talks. This was particularly irresponsible, the company claimed, "given the continuing widespread uncertainty regarding the impact of human activity on potential global climate change." With that statement Peterson was contradicting his company's own researchers, one of whom, Ken Croasdale, had stated during a conference in 1991 that greenhouse gases in the atmosphere were rising "due to the burning of fossil fuels" and that "nobody disputes this fact."

Peterson framed his opposition to climate policy in noble-sounding terms. "Continued economic growth is essential to fulfilling the individual and collective hopes of all Canadians for a rising standard of living, and history reveals that an expanding economy is directly related to the efficient use of affordable energy," he argued in a letter to shareholders. "Yet the Government of Canada appears to be on a course to mandate a substantial reduction in the use of oil and other hydrocarbon fuels, based on the unproven hypothesis that consumption of these fuels will cause global climate change."

The Imperial CEO had a direct financial incentive to stir up doubt, confusion, and uncertainty about the scientific consensus to which his own company's researchers had contributed. On the eve of the Kyoto talks, Imperial was expecting oil sands production at its Cold Lake bitumen operation to grow from 73,000 barrels to more than 130,000 barrels per day. Meanwhile, production at its Syncrude plant was up to 201,000 barrels per day, averaging about 70 million barrels per year. Imperial was planning a $475 million expansion that could bring that annual total to 81 million barrels. "Bitumen production will become increasingly important to Imperial," the company said.

Peterson warned shareholders and allies of the company that these positive developments could be quickly reversed if Canada moved to restrict greenhouse gas emissions, as it had promised to do under Kyoto. "Apprise yourself of the facts, and make your views known to government in the months ahead," he said.

During Peterson's helm at the company, Canada's greenhouse gas emissions had soared, growing 13 percent from 1990 to 1998. And that national average actually understated the problem. The fossil fuel industry's own emissions grew nearly 30 percent over the same time period, largely due to expansions in the oil sands. But telling the truth about Imperial's contribution to the climate crisis wasn't in the company's financial interest, so Peterson lied instead. "One thing is clear in this debate. There is absolutely no agreement among climatologists on whether or not the planet is getting warmer or, if it is, on whether the warming is the result of man-made factors or natural variations in the climate," he wrote. "I feel very safe in saying that the view that burning fossil fuels will result in global climate change remains an unproved hypothesis."

Peterson was especially outspoken among oil sands CEOs. In a 2002 interview he called Kyoto "the dumbest-assed thing I've heard in a long time." But even the companies presenting a more moderate image agreed in private that Kyoto had to be weakened and ultimately destroyed. Around the time of the Ottawa denial event, Suncor, the descendant of oil sands pioneer Sun Oil, explained to the *New York Times* that meeting the protocol's targets would be no problem because the company had already cut emissions significantly and planned to spend $100 million on renewable energy projects by 2005. "We'll align with whatever governments determine to be the public interest," said Gordon Lambert, Suncor's vice president for sustainable development. Yet the oil sands producer merely outsourced its Kyoto opposition to others.

Earlier that year, two industry groups that Suncor belonged to—the Canadian Chamber of Commerce and the Canadian Association of Petroleum Producers—held a press conference in Ottawa making the hyperbolic claim that meeting the climate goals of Kyoto would cause $30 billion worth of damage to the national economy. "At this time, we believe that it would be foolish for Canada to ratify Kyoto without a clear understanding of the issues on a national basis," one speaker said. Those two organizations were meanwhile spending $225,000 a week on ads telling Canadians to "ask your MP to stop the rush to ratify," as part of a fake citizens group called the Canadian Coalition for Responsible Environmental Solutions.

Publicly, these attempts by Suncor, Imperial, and their industry groups to kill Kyoto were unsuccessful. The Canadian government ratified the treaty on December 17, 2002. But privately, the noise, doubt, and confusion they'd created around addressing climate change resulted in major concessions. The day after ratification, Liberal natural resources minister Herb Dhaliwal sent a letter to the Canadian Association of Petroleum Producers promising that oil sands producers and other fossil fuel companies wouldn't have to pay any more than $15 per ton to offset their carbon emissions, meaning that costs per barrel might only go up by 10 or 20 cents. And, the letter said, those producers could rest assured that the total emissions Canada would actually reduce under the protocol would be capped at a level deemed reasonable to the industry.

The oil sands industry was thrilled. "They understand the issue," one executive said of concessions made by the federal government. "It's manageable," concurred Lambert from Suncor, as the company's revenue looked set to hit a record high of $5 billion. To some in the government it was hard to shake the feeling that policymakers—and the Canadian public—had been manipulated by a disinformation campaign whose ultimate goal had been to extract the best possible deal for oil producers.

"We've been subject to—and I am using my terms advisedly—a con game," said Liberal environment minister David Anderson.

IV

A Public Awakening

(1997–2008)

"Victory will be achieved"

IN THE SPRING OF 1997, the chairman of British Petroleum, one of the world's largest oil companies, gave an unexpected speech at Stanford University. "The time to consider the policy dimensions of climate change is not when the link between greenhouse gases and climate change is conclusively proven, but when the possibility cannot be discounted and is taken seriously by the society of which we are part. We in BP have reached that point," John Browne said. Shortly after, BP withdrew from the Global Climate Coalition. The oil and gas company was followed two years later by the automaker Ford. Then, in early 2000, the trickle of defections became a roar, with one corporate member after another rushing to publicly repudiate the once-powerful group. "Maybe it is time to ask the last one out to turn out the lights," the Sierra Club said.

During the 1990s, the Global Climate Coalition had transformed a bipartisan consensus about addressing climate change into a politically charged controversy, with many Americans no longer even believing the crisis was real. By 2000, the George W. Bush administration was explicitly acknowledging the coalition's influence in killing U.S. participation in the Kyoto Protocol. So why did the once-mighty Global Climate Coalition suddenly implode at the height of its political power? One explanation is that its members were terrified of being sued.

During the same years that Chrysler, Texaco, Shell, General Motors, and others were fleeing, the tobacco industry was losing the biggest corporate fraud lawsuit in history. Starting in 2000, cigarette makers were severely restricted from advertising their products and forced to pay over $206 billion in damages after being found guilty in court of lying to the public about whether smoking causes cancer. Oil companies and other climate polluters were watching closely. Writing about the Global Climate Coalition's collapse, journalist Lester Brown observed that "the high price paid by the tobacco industry's continuing denial of a link between smoking and health [was] all too familiar."

But the connections between climate denial and cancer denial weren't just metaphorical. During the 1990s, both campaigns were actually being coordinated by the same organization, a trade group known as the National Association of Manufacturers, which represented both oil refiners and cigarette companies. For much of its existence, the Global Climate Coalition was based out of the National Association of Manufacturers' Washington, D.C., headquarters. Documents show that the association was also helping Philip Morris and other tobacco companies undermine science that demonstrated a link between smoking and cancer.

All the strands came together in a group called the Advancement of Sound Science Coalition. Formed in 1993 at the behest of Philip Morris, it was a neutral-sounding organization that led media attacks attempting to discredit health warnings about secondhand smoke in restaurants and other public venues. But it soon grew into a one-stop shop for any corporate backer looking to dismiss scientific findings that were bad for their business model. Inevitably, that included climate denial. One of the organization's advisers was the climate contrarian Patrick Michaels. And the Advancement of Sound Science Coalition was coauthor along with Exxon and the American Petroleum Institute of the notorious 1998 memo referenced in the introduction to this book that explained that "victory will be achieved when... average citizens 'understand' (recognize) uncertainties in climate science."

That same memo referenced survey results showing that many Americans "currently perceive climate change to be a great threat." But it explained that "public opinion is open to change on climate science." When survey respondents were informed that "some scientists" doubted that humans were causing the crisis, a majority of those respondents began expressing doubt about whether climate regulations were a good idea. By 2000, however, it was becoming obvious to some climate polluters that public opinion wasn't malleable in only one direction. When people had learned in court about the disinformation campaigns of smoking companies, support for those companies evaporated and courts ordered them to pay hundreds of billions of dollars in fines. What had once been one of America's most powerful industries suffered a blow from which it never fully recovered.

"They lied about everything"

FOR AS LONG AS he could recall, Steve Berman disliked the tobacco industry. "My mother was a smoker," he said. As a child he vividly remembers "listening to her cough and hack all the time, she couldn't stop herself." It instilled in him a sense of injustice at the way some corporations operate. "I was just angry at the tobacco companies for putting this product on the market to begin with," he said.

Berman's anger wasn't misplaced. For decades, the big cigarette makers had used every corporate trick to keep people addicted to their product and avoid being held responsible for the death and sickness it caused. They attacked the science linking cigarettes to cancer, lobbied policymakers relentlessly to prevent profit-harming restrictions on smoking, and shut down one lawsuit after another with teams of aggressive and highly paid lawyers. But in 1994 the tobacco companies suddenly started to look vulnerable to legal action. And Berman saw an opportunity to go after them.

That was the year that Mississippi attorney general Mike Moore filed litigation against Philip Morris, R. J. Reynolds, Altria, British American Tobacco, and nine other top cigarette companies for knowingly sickening the state's citizens and burdening Mississippi's health care system with the costs. Though few predicted it at the time, this legal action would eventually bring Big Tobacco to its knees and result in one of the largest corporate settlements in history.

Berman grew up in Highland Park, an upper-middle-class community north of Chicago, and initially he felt little interest in the law. *To Kill a Mockingbird,* the seminal novel about lawyer Atticus Finch defending Tom Robinson, a Black man falsely accused of raping a white girl, came out in 1960, when Berman was still a kid, and inspired many attorneys of his generation, but Berman says it didn't spark his interest. Still, he can vividly recall the progressive ethos of an era when racial and social disparities were being widely questioned and challenged across society. His teen years were marked by the reactionary chaos that accompanied social progress: the assassinations of Martin Luther King Jr., Malcolm X, John F. Kennedy, and Robert Kennedy; the election of Richard Nixon; and anti–Vietnam War demonstrators at the 1968 Democratic National Convention being beaten by cops and the National Guard in the streets of Chicago—a four-day battle triggered by an aggressive police response that resulted in 668 people being arrested and hundreds of protesters and officers injured.

"It was a really evolutionary time," Berman said. "I grew up with a mother who was one of the first women to burn her bra. I remember she came home one day and said, 'I got a job, I'm not doing any housework or cleaning, it's up to you guys.'" Berman's dad had originally wanted a career as an army officer and was "pro-military" for much of the 1960s. "When we got into the Vietnam entanglement, he was in favor, but by 1968 he thought it was terrible," Berman said. Disillusioned by the war, his dad ended up running an insurance company. "He kind of became liberal himself," Berman said.

What Berman remembers most from high school is trying not to get flattened on the football field. "I was playing [football], second-string quarterback, going into my junior year," he recalled. "I'm calling plays in the very first scrimmage. There's a guy across the way. He just hated me. I think it was anti-Semitic. He says, 'Berman, I'm going to fucking kick your ass every day.' He was an animal. All of a sudden I realized, I'm only 5′10″, and I'm not going to get any bigger. This guy is going to kill me. So I go, 'Time out.' I hand the helmet to the coach, take off my pads, run over to the soccer field. I tell the coach there, 'I want to play soccer.'"

Berman's pragmatism was evident in other areas of his life. He knew some people in Highland Park who went from progressive to radical. They set off bombs and had to run from the law. "They were elder brothers of some of my close friends and I've never seen them again," he said. Were they associated with the Weathermen, the left-wing organization active in the U.S. Midwest at the time, which bombed government buildings and banks and broke Timothy Leary out of jail? Berman isn't sure, saying only that "they were in that genre of people out there doing radical stuff."

That experience imparted an important life lesson. "I said, 'That's really not the way to make change,'" he remembered. "The way to make change is to get trained in what the establishment does, and make change through the process."

In the 1970s, he enrolled at the University of Michigan. Berman's shift into law came during his senior year of college. "It's a very sad story," he later explained. "I was going out with a woman who was in law school who passed away, and I think that I wanted to finish it for her. It took me years to psychoanalyze how I got to law school from that."

Berman moved out to Seattle after graduating, where he got hired by a firm that mostly did defense work for companies in industries like insurance. Berman was part of a smaller team at the firm that sued companies instead. He brought legal action against so many companies in those early days—including Nordstrom, Weyerhaeuser, SeaTac Airport, Microsoft, and Boeing—that he eventually earned the nickname "Berman the vermin."

The case that would truly launch his career began in 1993. Early that year, hospitals throughout the state of Washington noticed a disturbing pattern. Hundreds of patients were showing up at emergency rooms experiencing bloody diarrhea. Many of them were children, who upon examination were diagnosed with hemolytic uremic syndrome, a condition in which blood vessels in the kidneys are damaged and inflamed. The Washington State Department of Health conducted an investigation and found a common link: nearly all of the patients had eaten hamburgers at the fast-food chain Jack in the Box shortly before becoming seriously sick.

Around this time Berman met with a parent whose child had just lost a kidney. The parent wanted to sue Jack in the Box. "I remember sitting in the room thinking, 'This could have been my kid,'" Berman said. "I was so choked up I could barely talk." Berman thought there was a strong case against the company, but his employer disagreed. The firm worried that the lawsuit Berman was proposing could cause them to lose an insurance client. Berman pushed back. "Look, this is why I became a lawyer," he said. His superiors were unmoved. "So I said, 'I'm out of here.'"

Berman teamed up with a colleague named Carl Hagens, and together they started their own firm, Hagens Berman. He'd wanted to set out on his own for some time, and here was the perfect excuse. Two years later they settled a shareholders' case with Jack in the Box for $12 million.

But that case was tiny for Berman compared to what would come next. Berman first heard about the Mississippi legal effort against

Big Tobacco in 1994. As Berman read through the case brought by Attorney General Moore, he realized there was an opportunity to fight—and potentially win. At the time there was very little interest among Berman's peers for taking on Big Tobacco. There was also concern within Berman's firm itself about whether joining the nascent state-led fight against cigarette companies was a good idea. The other partner in the firm, Carl Hagens, "made me promise I'd only dedicate a certain amount of time to the case because he was afraid we'd lose," Berman said. "If we devoted all our time to a losing effort and we didn't have any income, then we could go under." It was the mid-1990s, Berman explained, and "people have to remember that nobody had won a tobacco case."

The very first anti-smoking case went to trial in 1962. It was brought on behalf of a sixty-two-year-old businessman in Missouri who claimed in court that he'd lost his larynx due to smoking three packs a day. The legal team for Philip Morris flooded the court with health experts claiming that there wasn't enough evidence to link tobacco smoke to head and neck cancer. The jury only deliberated for an hour before ruling against the Missouri businessman. That set a precedent for dozens of lawsuits to come: tobacco companies amplified uncertainties in the science of cancer to create doubt in the minds of jurors and judges. Over time they added another layer to their defense. Since cigarette labels contained health warnings, the companies argued, people who smoked knew the harms, and therefore the companies couldn't be held liable. "Tobacco did a really good job of saying you can't prove it causes cancer or else saying 'It's your own damn fault,'" Berman said.

This denial strategy reached an infamous crescendo on April 15, 1994, when the heads of the world's top seven tobacco companies testified to the U.S. Congress that cigarettes aren't addictive nor a significant cause of cancer. They made these statements despite their companies having conducted years of internal research proving otherwise. (A tobacco chemist with R. J. Reynolds had acknowledged the "overwhelming" evidence proving smoking's link to lung cancer back in 1962.) At one point during the congressional hearing, Democratic representative Henry Waxman asked Andrew Tisch, the chairman and chief executive of the Lorillard Tobacco Company, to comment on whether cigarette smoke is a factor in cancer. "I do not believe that," Tisch said. Waxman pushed back: "Do you understand how isolated you are from the scientific community in your belief?" The tobacco executive replied, "I do, sir."

But by the time Berman got involved, this denial strategy was starting to fall apart for tobacco companies. In the mid-1990s, a whistleblower went public with information that companies had put addictive chemicals into cigarettes that increased the risk of cancer. Jeffrey Wigand was a lead researcher for Brown & Williamson, the third-largest tobacco company in the United States, when he discovered that a cigarette additive called coumarin had similar properties to rat poison and caused tumors in mice. After Wigand informed the company of this, they told him that removing coumarin from its products could hurt sales. The company eventually fired Wigand and hired private investigators to find dirt on him. In 1995, he gave a deposition explaining how the tobacco industry was lying to the public and to what lengths it would go to silence critics. That deposition was then used in the lawsuit brought by Mississippi's attorney general.

Meanwhile, documents began to leak from tobacco companies showing what executives knew about addiction and disease and their strategies for manipulating the public. One collection of eighty-one papers included "confidential marketing surveys, communications to the [R. J. Reynolds] board of directors, reports by outside advertising firms, long-term planning documents and other internal memos, all of which deal with the youth smoking market," as the *Washington Post* reported at the time. One paper quoted the company's marketing vice president, C. A. Tucker, who talked about the need to turn teenagers into smokers. "They represent tomorrow's cigarette business," Tucker said. "Our strategy becomes clear for our established brands... Direct advertising appeal to the younger smokers."

The involvement of attorneys general like Mike Moore also contributed to a much stronger—and potentially winnable—case against the companies. Most of the previous legislation was brought on behalf of individuals. "It was the smoker versus the tobacco company," Berman said. But litigation brought by Mississippi, Washington, and eventually forty-four other states was different. "Our case was that the state of Washington is not a smoker," Berman said. "The state didn't choose to smoke and we could use epidemiological studies to show how much lung disease was caused because of smoking... we can do it through the statistics and say, 'Look, because you lied about your product, all these people are using it and we are paying the health care bill.'"

Like he had done with Jack in the Box, Berman threw himself into the case. Having started with Washington State, he was soon representing Arizona, Illinois, Indiana, New York, Alaska, Idaho, Ohio, Oregon, Nevada, Montana, Vermont, and Rhode Island. Not only was he fighting against some of the most formidable corporate giants in the United States, he also risked bankrupting his own firm. "The quiet, youthful [lawyer] worked brutal hours to make up for his firm's small size, arriving at his office at six A.M. and staying until nine or ten P.M. six days a week," Dan Zegart writes in *Civil Warriors: The Legal Siege on the Tobacco Industry.*

Within a few years, Berman's hard work was starting to pay off. Though the tobacco companies presented themselves publicly as a united front, behind the scenes a smaller cigarette maker known as Liggett began to worry about lawsuits filed by the attorneys general. If even one of the state-led cases was successful, the company feared, it could easily go bankrupt. Privately, Liggett began meeting with litigators about a settlement. Berman was in contact with Liggett's CEO during this period, and he helped convince that tobacco executive to come out as a whistleblower against the other companies. On March 20, 1997, Liggett publicly agreed to pay $750 million while admitting that cigarettes cause cancer. "This is the beginning of the end for this conspiracy of lies and deception," Arizona attorney general Grant Woods said at the time.

Berman gave the opening statement at the Washington trial. There were hundreds of people gathered to watch. All the major news networks were present with cameras running. Standing inside the Seattle courthouse, Berman went on the attack. "I had a big poster that said, 'The Industry's Five Big Lies,' and I went through each one of those lies," he recalled. Berman dismantled decades of tobacco talking points during his argument: that smoking doesn't cause serious disease, that it isn't addictive, that nicotine wasn't strengthened in tobacco labs, that "independent research" showed no health effects, and that the industry wasn't interested in marketing to children. "It was obviously quite an exciting moment," he said.

By then, however, the industry was ready to surrender. In November 1998, Big Tobacco officially agreed to pay a total of $206 billion to forty-six states. As part of that agreement the industry received immunity from other pending lawsuits similar to the ones led by Moore and Berman. Yet it was still a massive victory, the largest corporate settlement in history. "We were pretty damn happy," Berman said. Speaking to reporters, Washington State attorney general Christine Gregoire said it marked the end of a deceitful era: "Joe Camel and his ilk are now in intensive care and will be gone by April. Billboards will be coming down, and the real truth about tobacco will be available to every American."

Berman was at the press conference too. He described the tobacco settlement as a victory for truth. "They lied about everything," he said. The case turned him into a legal superstar, demonstrating that even the most powerful industries could be taken down once their protective lies were deconstructed from within.

It wouldn't be long before Berman would attempt to use the same legal strategy against the oil industry.

"Saudi Arabia of the western world"

IN AUGUST 2003, public health authorities broke down the door of a tiny attic apartment in Paris. They'd been alerted to a terrible smell and inside they found its source: a seventy-seven-year-old Serbian man named Petar who'd been dead for at least two weeks. Alone and without air conditioning, he'd become dehydrated and disoriented during a heat wave that resulted in weeks of temperatures above ninety-five degrees, eventually collapsing and dying. Petar was one of thousands to die across France that month. "The tally continued to rise as vacationers returned home to apartments reeking of decay, with ceilings and walls fouled by the body fluids of those who had died in attic apartments," according to one account. By the end of the summer, the heat wave had contributed to the deaths of up to seventy thousand people.

Scientists would later determine that this heat wave was made less survivable by climate change. Atmospheric concentrations of carbon had hit 375 parts per million, a 33 percent increase since the early 1800s. Heat waves are common across Europe. But this one was particularly unbearable—the extra length and intensity caused by climate change was responsible for an additional 506 out of 735 heat deaths in Paris alone, University of Oxford climate scientist Daniel Mitchell calculated. "People can really start to understand that these are impacts we're seeing now, not in the future," he said.

But in Alberta's oil industry, all anyone could talk about that summer was a much different statistic. After decades of the oil sands being viewed as an esoteric and isolated corner of the global oil market, producing what Koch Industries had referred to as "garbage crudes," the industry had finally been recognized in the mainstream. In 2003, the analytical arm of the U.S. Department of Energy, an influential organization known as the Energy Information Administration, officially acknowledged that the oil sands could hold 180 billion barrels of recoverable oil, placing Canada in a similar league to Saudi Arabia and Venezuela.

"Something very important has just happened," Daniel Yergin, a respected oil industry analyst and historian, said of the rising geopolitical importance of Alberta. That same year, global oil prices jumped to above $45 per barrel, suddenly enabling higher profits for oil sands projects that required a price of $23 per barrel just to break even. Now, investment was pouring into northern Alberta at record levels, and bitumen was pouring out. In 2004, Canada for the first time overtook Saudi Arabia as the largest exporter of oil to the United States. "No one disputes that the oil-sands industry has come of age in Canada," CNN reported. It was "a big deal that will only get bigger."

The actual operations were becoming gigantic. In 2003, Shell, Chevron, and a consortium of other producers opened up the Athabasca Oil Sands Project, a $3.7 billion mine and processing plant capable of pumping 160,000 barrels per day. At the mine site, shovels clawed into the boreal landscape, ripping a hundred tons of oil sands at a time from the ground and dumping it into heavy hauler trucks. The hauler trucks, equivalent in size to small buildings, then deposited their loads in a crusher, where big chunks of oily earth were smashed into smaller pieces, turned with warm water into a slurry, and processed into a frothy sludge inside an extraction plant. The waste ended up inside a series of vast polluted reservoirs that grew larger and more toxic each year. Added all up, those "tailings ponds" soon occupied an area larger than the city of Vancouver.

Meanwhile, Suncor had just completed building Project Millennium, a $3.4 billion expansion of the operation it had first begun running in 1967. This new megasite was soon pumping out 225,000 barrels per day, inspiring the company to conclude that "the future appears bright." On June 4, 2003, Imperial Oil officially opened a

$650 million expansion of its Cold Lake operation that would bring the area's production up to around 170,000 barrels per day. Imperial called it "a significant milestone," saying that it "incorporates the best of everything we've learned during nearly four decades of pilot work and commercial operations at Cold Lake."

Oil prices rose higher and higher: going from $45 to $60 to $80 to $90 within only a few years. By the middle of the decade there were over $80 billion worth of projects either underway or being planned in the oil sands. The *Wall Street Journal* described it as a "black-gold rush in Alberta," one in which "industry participants forecast that production from the oil-sands region will nearly triple to 2.7 million barrels a day by 2015 from one million barrels a day last year." With oil majors piling into the region, the *Financial Times* deemed it "North America's biggest resources boom since the Klondike gold rush." Writing in the *Nation*, Naomi Klein reported that "the town of Fort McMurray, ground zero of the boom, has nowhere to house the tens of thousands of new workers, and one company has built its own airstrip so it can fly in the people it needs."

The boom was also happening south of the border. By 2008, as oil prices soared as high as $147 per barrel, refinery owners across the United States spent $53 billion to process bitumen from the oil sands. That year, there were seventeen refinery expansions and five new refineries under construction or being considered, according to the Environmental Integrity Project, a nonprofit organization. All of this stood to raise the processing capacity for oil sands crude by more than one million barrels per day. "This data suggests that U.S. refineries are placing a major bet on a fuel source which is dirtier to mine, process and refine, and the extraction of which releases three times more greenhouse gas as conventional crude oil," a report from the organization warned.

Few refiners were as deeply committed to the oil sands as Koch Industries. Its Pine Bend Refinery, a massive complex covering one thousand acres with three thousand miles of pipes, was by the end of the decade turning 320,000 barrels of oil per day into gasoline and other climate-harming products. About 80 percent of that oil supply came from Alberta. Even though Koch Industries had been processing oil sands crude for decades, far longer than most other refiners, bitumen still presented difficulties for the company owing to its high sulfur content. "Its physical properties are denser, which make it more challenging to refine," Jake Reint, communications director for Pine Bend, told a local news outlet, calling the oil "very corrosive."

But the technical challenges of handling this gusher of heavy oil were very much worth it. "After decades, Pine Bend still benefited from occupying a stunningly profitable bottleneck in the U.S. energy system," Christopher Leonard writes in *Kochland*. "Cheap oil from Canada's tar sands piled up at the U.S. border without many buyers except Koch, and Koch could still sell its refined gasoline into midwestern markets where prices were relatively high thanks to a lack of refining capacity."

As all this took place, emissions from the oil sands were going into hyperdrive. The greenhouse gases emitted from new megaprojects caused the oil sands industry's climate footprint to reach fifty-five megatons by the early 2010s, a 224 percent increase from the early 1900s. This figure was more than the entire emissions footprint produced by Sweden, and nearly as large as the emissions released by the 150 million people living in Bangladesh. Even that doesn't fully capture the scale. "With a more comprehensive accounting of emissions, including those linked to upgrading exported bitumen in the US, production and processing of natural gas used in the oil sands, production of other required inputs, and loss of carbon storage due to the disruption of boreal forest, total GHG emissions could be as much as double the federal government's figures," wrote Anders Hayden, a professor at Dalhousie University.

At this rate, the Kyoto Protocol climate goals Canada had agreed to would be impossible to achieve. Others saw in the explosion of the oil sands a threat to the entire global climate system. A 2009 Greenpeace report argued that "Canada is now one of the world's leading emitters of [greenhouse gases], and a global defender of dirty fuels." Alberta's vast bitumen deposits, it concluded, had transformed America's polite northerly neighbor into "the Saudi Arabia of the western world."

"What Makes Weather?"

IN LATE 2007, one of the world's largest oil and gas companies announced it was for the first time entering the oil sands. BP was coming to the bitumen gold rush a bit late. Ever since the price of oil shot up to $100 per barrel in the 2000s, suddenly making the industry unthinkably profitable, international oil majors had been rushing to buy up deposits and assets in northern Alberta. BP was the only oil major that hadn't, owing to its CEO John Browne's concern that producing bitumen was too expensive. "He came to dislike the oil sands, and believed that it wasn't something that you should have in your business," Bill McCaffrey, CEO of the Canadian oil sands producer MEG Energy, explained. "It didn't fit into his strategic model. There are different thoughts and different strategies and at that time, for Sir John, it didn't fit into the BP portfolio."

But under the new leadership of Tony Hayward, who became chief executive in 2007, that thinking had radically changed. BP would now be joining with Husky Energy, a Calgary-based company run by Hong Kong billionaire Li Ka-shing, on a $5.5 billion project called Sunrise. They hoped to tap a rich bitumen deposit located around forty miles northeast of Fort McMurray, containing potentially 3.7 billion barrels of oil. "The huge potential of Sunrise is the ideal entry point for BP into Canadian oil sands," Hayward said in December 2007.

It was a massive international project. According to the proposal, the two companies would produce 200,000 barrels per day of bitumen using a technology called steam-assisted gravity drainage, a process by which they'd pump steam deep into the earth, heating the tarry bitumen into a smoother-flowing state, and then suck it up to the surface. From there, the bitumen would be sent via pipeline to Hardisty, Alberta, and then piped south of the border. Some of the oil would end up in BP's refinery in Toledo, Ohio, which the company was expanding so that it could process up to 170,000 barrels of heavy oil per day.

"The result will be the development of a major new Canadian oil field and the modernization and expansion of the Toledo refinery to allow far greater use of Canadian heavy oil," BP America chairman and president Bob Malone said at the time. BP and Husky expected to produce oil for at least four decades. But there was just one problem with this plan. Bitumen was some of the most carbon-heavy oil in the world, resulting in emissions anywhere from 14 to 20 percent higher than regular barrels of crude oil, according to a U.S. Congressional Research Service report. The same chemical properties that made bitumen unappealingly expensive for BP's former executive Browne also made it terrible for the climate.

Tapping these huge deposits would therefore exacerbate what BP had identified years earlier as "one of our urgent concerns"—the warming of the atmosphere due to burning fossil fuels.

Like Shell, Suncor, and Imperial Oil, which by this point had operated in the oil sands for several decades, BP knew about the dangers of climate change long before its executives acknowledged them publicly. This is evident from a twenty-four-minute documentary released in 1991 by BP's Corporate Communication Services called *What Makes Weather?*

Over stock footage of mountains, polar research stations, ocean marine life, and busy streets in London, a British-accented narrator explained that "Our whole energy-intensive way of life and its dependence on carbon-based fuels is now a cause for concern." The reason for this, he said, is "when coal, oil, or gas are burned, they release carbon dioxide and other reactive gases. Since the industrial revolution, their use has increased hundredfold. In the last forty years, the mass burning of the tropical forests has freed even more carbon dioxide into the atmosphere. It has taken time to realize what damage this extra carbon dioxide can do."

BP was explicit about the "devastating consequences" in store unless global emissions were brought under control. "If the oceans got warmer and the ice sheets began to melt, sea levels would rise, encroaching on coastal lowlands. From warmer seas more water would evaporate, making storms and the havoc they cause more frequent... and low-lying countries like Bangladesh would be defenseless against them," the narrator said. "Away from the coasts we could see a return to the conditions that devastated America's Midwest in the 1930s. Global warming could repeat on a more disastrous scale the dust bowl phenomenon, which virtually destroyed farming on the great plains." Because of this, the BP-produced film concluded, climate change is "one of our urgent concerns."

The film was likely viewed by "hundreds if not thousands" of people, according to the European media outlet Follow the Money. BP since the 1950s had produced dozens of films for educational purposes and appears to have sold *What Makes Weather?* through its education service for a price of £12.95 throughout the 1990s. The documentary was even submitted to the thirty-ninth annual Columbus International Film and Video Festival, where it won a bronze plaque. But the film's dire warnings about the damage BP's business model was doing to the climate were not apparently heeded by the company's leadership.

The same year that *What Makes Weather?* came out, BP America was included on a membership list for the Global Climate Coalition, the organization created by fossil fuel, vehicle, manufacturing, and other high-emitting corporations in the late 1980s to spread doubt and uncertainty about climate science. Reports released by the coalition directly contradicted the warnings contained in BP's documentary. For instance, a 1994 report from the coalition titled "Potential Global Climate Change" stated that "there is no evidence of a warming trend that can be traced to man-made emissions." The report also argued that "claims of an impending catastrophe rest on little more than speculation."

By 1997, however, BP was ready to defect from the coalition. Its then CEO Browne had apparently decided it was better for the oil company's interests to acknowledge that climate change exists and position itself as part of the solution. "I was the chief executive of an oil company and I was about to become an environmental activist," Browne later wrote in his memoir. During a May speech at Stanford that year, referenced a few chapters earlier, he became the first major oil executive to state that climate change is real. "There is now an effective consensus among the world's leading scientists and serious and well informed people outside the scientific community that there is a discernible human influence on the climate and a link between the concentration of carbon dioxide and the increase in temperature," he said.

Browne also revealed BP had been calculating its own contribution to climate change. "We've looked carefully using the best available data at the precise impact of our own activities," he said. That data showed BP's impact on the climate to be around ninety-five megatons per year worth of carbon emissions. However, Browne was quick to downplay that figure. "Our contribution is small," he said of an atmospheric impact greater than the entire carbon footprint of the city of New York. The BP executive also ruled out climate solutions that would impact the company's business model. "Actions which sought, at a stroke, drastically to restrict carbon emissions or even to ban the use of fossil fuels would be unsustainable because they would crash into the realities of economic growth," he said.

That speech, combined with BP's subsequent resignation from the Global Climate Coalition, resulted in tons of positive publicity.

"On balance, BP is probably the environmentalists' preferred oil company, while Mr. Browne is definitely their favourite oil boss," Geoffrey Lean wrote in the *Independent*. Despite the progressive optics, BP's behavior hadn't really changed at all. It still remained a member of the American Legislative Exchange Council, a group that ran disinformation campaigns on climate change alongside the Global Climate Coalition, and that by 1999 had helped pass regulations blocking restrictions on greenhouse gases in Alabama, Colorado, Indiana, Ohio, and a dozen other U.S. states. "Carbon dioxide, the inescapable by-product of burning fossil fuels, is beneficial to plant and human life alike. The effort to regulate it as a greenhouse gas is an attempt to tax energy," a spokesperson for the American Legislative Exchange Council said in 2004.

Meanwhile BP's oil and gas production kept expanding. By 2006, it was extracting oil from twenty-two countries and exploring for future reserves in the Gulf of Mexico, Angola, Egypt, Russia, and Algeria. The following year, BP announced its $5.5 billion expansion into the oil sands. Li, the Hong Kong billionaire whose company Husky owned half of the proposed Sunrise operation, called it a "mega oil sands project." Greenpeace responded to the deal by accusing BP of committing a "climate crime." But the company, which as early as 1991 had warned that global temperature rise could threaten the lives of millions of people in Bangladesh and other low-lying coastal regions while also destabilizing food production on a global scale, was unrepentant. "Someone is going to develop those resources," a BP spokesperson said.

"Global energy powerhouse"

STÉPHANE DION looked uncomfortable. The bespectacled leader of Canada's Liberal Party was waging a national election campaign at the height of the financial crash in autumn 2008, and this day he was on television to share his thoughts about protecting Canada from the cascading crisis. "Can we start again?" he asked the CTV News interviewer. Dion, whose first language is French, was having trouble understanding the ambiguous verb tense of a question about the crash. "If you were prime minister now, what would you have done about the economy and this crisis that [the current prime minister] has not done?" CTV anchor Steve Murphy asked.

"I don't understand the question, sir," Dion replied. "Are you asking me to explain at which moment, today, or since a week, or since two weeks?"

"No," Murphy said, "if you were the prime minister during this time already." Dion looked flustered and faced the camera. "We need to start again," he said.

To many observers of the 2008 Canadian federal election, this was a key moment when Dion lost control of the race. Dion's campaign was furious that CTV decided to air the footage of the Liberal leader stuttering through a simple question, and during the ensuing controversy media pundits debated the ethics of the decision. "We decided that it was important that CTV News not hide anything during an election campaign," the network's president, Robert Hurst, said in defense. By then the damage had been done: Dion looked like a weak and ineffectual communicator during a global emergency that demanded unwavering leadership.

The irony was that Dion had based his campaign around one of the strongest climate plans proposed to date anywhere in the world. While opponents portrayed him as a stuttering egghead, Dion was prepping for battle against the world's largest oil companies, which at that point were earning unprecedented fortunes in the oil sands. Dion's "Green Shift" was not so different from the carbon pricing policy that Imperial Oil had modeled seventeen years earlier. If elected, Dion had promised to make polluters pay $10 for every ton of carbon dioxide they emitted into the atmosphere, a tax that within four years would quadruple to $40 per ton. Since that would result in ordinary people paying higher prices for gasoline and heating fuels, Dion said some of the government revenue raised from the tax would be used to lower people's income taxes. "Our plan will be good for the environment and good for the economy—good for the planet, good for the wallet," he said.

But awkward TV interviews weren't Dion's only problem. His main opponent during the election was Stephen Harper, a Conservative candidate who at that point had been prime minister for two years. Harper was ruthless in his attacks on Dion's climate plan. "This will actually screw everybody across the country," Harper said. He seized on the fact that gasoline prices were already at record highs due to the global price of oil going above $100 per barrel. "We think that is a foolish and unnecessary policy," Harper said of the Green Shift. Harper's attacks, combined with doubts about Dion's ability to communicate effectively, resulted in a crushing defeat on election day. Dion received only 26 percent of the popular vote and Harper was easily reelected.

The election was seen as a cautionary tale, not just in Canada, for any national leader attempting to propose economy-altering climate legislation. But it also clearly demonstrated the vast and growing political power of Canada's heavy oil. Unlike with the battles over the Kyoto Protocol, producers such as Shell, Suncor, and Imperial had not needed to wage a coast-to-coast climate disinformation offensive against Dion's Green Shift. That's because they had direct access to a much more powerful weapon: the office of the prime minister himself. Oil executives didn't need to win Harper over to their cause because Harper was already on their side.

Harper had spent much of his life in close proximity to the oil sands. Born in Toronto, he moved to Alberta in 1978 to take a desk job with Imperial Oil, which had also employed his father. Harper began working at the Exxon subsidiary just as it was starting to produce bitumen from the province's north. Harper later studied economics at the University of Calgary, where the school's political scientists were at the time largely recruited from conservative networks in the United States. Harper's political views were formed by reading the libertarian canon: Friedrich Hayek and William Buckley "left him with a profound respect for the workings of a free-market economy and a set of neo-conservative beliefs that were being put into political practice by Margaret Thatcher in Britain and Ronald Reagan in the United States," according to one account.

Harper was not someone who appeared destined for a career in politics. He spoke in flat monotone sentences and could smile with his mouth but not so much with his eyes. But where he lacked charisma, Harper had a "cold brilliance and a cold arrogance," as the outlet *iPolitics* described him. He had the cool demeanor of a strategist with an ability to spot emerging political trends. During the early 1990s, Harper and his fellow conservatives in Alberta could see U.S. politics swinging hard to the right. Grover Norquist was gaining influence within the Republican Party for articulating a strain of anti-tax populism around the idea that government should be shrunk "down to the size we can drown it in a bathtub." And a grassroots evangelical network known as the Christian Coalition was helping Republicans sweep midterm elections.

After spending a decade dabbling in various experiments to export this populist conservatism to Canada, Harper became leader of a political party called the Canadian Alliance in 2002. The party's views were far outside the mainstream. Previous leader Stockwell Day had argued that same-sex marriage was sinful and was mocked relentlessly in the media for his belief that the world was only six thousand years old and that humans lived at the same time as the dinosaurs. Harper himself wasn't all that religious, and in his early days as leader of the Canadian Alliance he was distrusted by the party's evangelical members. In order to win real political power for the party, Harper knew that he had to unite the Alliance's fiscal libertarians with its large faction of socially conservative Christians.

One way to achieve this was to channel the ideological forces that had made oil sands development possible generations earlier. During the 1960s, Sun Oil chairman J. Howard Pew had bonded with Alberta premier Ernest Manning over their deep evangelical beliefs and free-market hatred of communism. That friendship played a crucial role in Manning removing regulatory barriers and allowing Sun Oil, an American company, to build the first commercial bitumen operation. Echoes of that union could be heard in a 2002 fundraising letter that Harper sent to Alliance supporters. "We're gearing up now for the biggest struggle our party has faced since you entrusted me with the leadership," he wrote. Harper was referring to Canada's ratification of the Kyoto Protocol, "a socialist scheme to suck money out of wealth-producing nations." Even worse, he argued, was that the "job-killing, economy-destroying" protocol was being used to limit carbon dioxide, a gas that he described as "essential to life."

The born-again Christians and Hayek-worshipping tax haters in Harper's party might disagree on a lot. But Harper correctly bet that they could be united in defense of the country's oil and gas sector. Over the following years, Harper skillfully incorporated the concerns of his unwieldy coalition into a coherent political agenda that included lower taxes, tougher penalties for crime, opposition to same-sex marriage, distrust of climate regulations, and closer relations with the United States. When Canada under the Liberal government of Jean Chrétien refrained from joining George W. Bush's disastrous invasion of Iraq, Harper penned an open letter in the *Wall Street Journal* calling the decision a mistake. "Disarming Iraq is necessary for the long-term security of the world," he wrote.

By 2006, the Alliance had merged forces with a reinvigorated Conservative Party. And with Harper at its helm, it cruised to victory in that year's federal election. "Tonight, friends, our great country has voted for change," Harper said during his acceptance speech. Canada, the BBC wrote, "has swung to the right." But there was more to Harper's victory than simply a lurch in political values. In some ways it was the culmination of a half century of climate disinformation. Harper's former employer Imperial Oil had produced clear science about the dangers of burning fossil fuels, and then spread denial once it was clear that stopping global temperature rise would harm its business model. The company's strategy for defending the oil sands in the midst of an escalating climate crisis was now official government policy under Prime Minister Harper. "It's a complicated subject that is evolving," he said in 2006 when asked about climate change. "We have difficulties in predicting the weather in one week or even tomorrow. Imagine in a few decades."

Professional climate change deniers noted these comments approvingly. Shortly after Harper's election win, Patrick Michaels and several dozen other contrarians wrote a letter to the new Conservative prime minister. "The study of global climate change is, as you have said, an 'emerging science', one that is perhaps the most complex ever tackled," the letter notes. But Michaels and his cohort suggested Harper could be even stronger in his language. " 'Climate change is real' is a meaningless phrase used repeatedly by activists to convince the public that a climate catastrophe is looming and humanity is the cause. Neither of these fears is justified," they wrote. It's not known if Harper responded, but the following month, leaked documents from the international climate negotiations underway in Bonn, Germany, revealed a new diplomatic strategy for Canada: oppose all efforts to create "stringent targets" worldwide for reducing emissions.

This was crucial for Canada's oil sands, as Harper made clear that July during his first international speech. Addressing a crowd of three hundred Canadian and U.K. businesspeople in London, Harper referred to Canada as an "emerging energy superpower." The country, he said, ranks seventh in the world for oil production. "But that's just the beginning," he said. "An ocean of oil-soaked sand lies under the muskeg of Northern Alberta—my home province." At that time there were $100 billion worth of bitumen projects planned or under construction. Harper vowed that his government was going to transform Canada into a "global energy powerhouse."

Over the next two years, the former Imperial employee fought off anyone who would challenge that plan. He appointed right-wing climate deniers throughout the Canadian bureaucracy, announced a new policy preventing federal climate scientists from speaking to journalists, canceled billions of dollars of federal spending on green energy, embarked on an extensive government review of the Kyoto Protocol, and crushed Dion's 2008 election plan to bring in an oil-sands-penalizing carbon tax.

By the end of the year, bitumen production had reached a new high of 1.2 million barrels per day. But there was a growing political threat south of the border. The United States had just elected Barack Obama, and the president was eager to fix the climate emergency.

His chances of success seemed solid. In fact, the potential for a united legislative effort to deal with the coming catastrophe was so tangible that in 2008 Republican former House Speaker Newt Gingrich, long a staunch ally of oil, teamed up for a television ad with then Democrat House Speaker Nancy Pelosi, who was trying to push a cap-and-trade bill through Congress that had potential to dramatically lower U.S. emissions. The sworn enemies were filmed sitting together on a love seat. "We do agree our country must take action to address climate change," Gingrich pronounced in the ad, sponsored by former vice president Al Gore's Alliance for Climate Protection. "If enough of us demand action from our leaders, we can spark the innovation that we need."

Harper was determined to make sure that didn't happen.

V

Blame Canada

(2006–2010)

"Back off dudes!"

IT WAS A SUNNY JUNE DAY in Las Vegas and Barack Obama was full of optimism. The senator from Illinois had three weeks earlier been chosen by Democrats to compete for president in the 2008 U.S. election. Obama had made it clear that stopping climate change was going to be a significant part of his campaign. "This is the moment when the rise of the oceans began to slow and our planet began to heal," he famously said during his acceptance speech for the Democratic nomination. Throughout the month, Obama's Republican competitor John McCain had been doing media events talking about his plan to expand offshore oil drilling. Obama headed to Nevada to present an energy plan of his own.

Trailed by cameras, the first-ever Black presidential contender was given a tour through Springs Preserve, a 180-acre botanical garden and museum not far from the Las Vegas strip featuring exhibitions about sustainability. "What we are seeing here—from the solar panels that power this facility to the Bombard workers who built it—is that a green, renewable energy economy isn't some pie-in-the-sky, far-off future, it is now," Obama said in a speech. "It is creating jobs, now. It is providing cheap alternatives to $140-a-barrel oil, now. And it can create millions of additional jobs and entire new industries if we act now." In contrast to McCain, who promised to tap 21 billion barrels of oil buried deep in the ocean floor by lifting a federal offshore drilling ban, Obama declared petroleum the enemy. "Each and every year," he said, "we become more, not less, addicted to oil—a nineteenth-century fossil fuel that is dirty, dwindling, and dangerously expensive."

Obama had not named the oil sands directly, but his speech set off alarms in Canada, where his statements referring to oil were carefully parsed by the industry's defenders. "The word 'dirty' is a dead giveaway," conservative columnist Peter Foster wrote in the *National Post*. "Mr. Obama, speaking in Las Vegas, implied that the U.S. might somehow do without the oil sands."

With momentum and excitement building behind Obama's campaign for president, Canada's federal government knew this was a threat it couldn't ignore. It scheduled a meeting with the Obama campaign later that summer. Over breakfast in Washington, D.C., Obama's top energy adviser, Jason Grumet, met with Tony Clement, a Canadian cabinet minister and trusted ally of prime minister Stephen Harper. They were joined in the meeting by executives from major oil sands producers. "We've got to really have a sophisticated full-court press on Canada's issues with the United States of America, and that's certainly what the prime minister is interested in," Clement explained.

Grumet was picked as a target of that "full-court press" for stating that greenhouse gas emissions from the oil sands are "unacceptably high." During an interview around the time of Obama's Nevada speech, Grumet said that "if the only way to produce those [oil sands] would be at a significant penalty to climate change, then we don't believe that those resources are... going to play a growing role in the long-term future." Over one million barrels per day of oil sands crude was flowing into the United States by that point, and industry analysts expected that figure to more than double by 2020. "If you don't like the oil sands oil, what companies [in Canada] will do is build a bigger pipeline to the west coast and export it to China and India," a lobbyist for the oil and gas producer Nexen warned the Obama campaign.

But that tough talk was masking real vulnerabilities. It's true that at that moment a Calgary-based pipeline company named Enbridge was proposing to build a 525,000-barrel-per-day oil sands pipeline to the West Coast. However, that project wasn't anywhere close to being constructed and was already causing outrage and opposition among First Nations along its proposed route. In one incident, Enbridge contractors accidentally sawed down fourteen old-growth trees in northwest British Columbia with ancient Haisla markings on them. "We compared it to a thief breaking into your house and destroying one of your prized possessions, and then calling you later to apologize for it," Haisla councilor Russell Ross Jr. explained at the time.

In the meantime, U.S. policymakers were proposing climate legislation that could harshly penalize the oil sands. California had recently adopted something called a low-carbon fuel standard, a policy that required the state to restrict the sale of gasoline and other fuels derived from oil sources that have a higher-than-average climate footprint. This would restrict any road fuel made from oil sands crude, because producing a barrel of bitumen can emit up to 20 percent more greenhouse gases than a barrel of conventional oil. Harper's administration intervened at least five times to block the California policy. But then Republican governor Arnold Schwarzenegger was unpersuaded. The standard, he said, "would dramatically increase low-carbon fuels, expand consumer choice, and reward innovation."

Obama himself had expressed support for taking the low-carbon fuel standard national during the early days of his presidential campaign. While visiting a gas station in the Los Angeles neighborhood of Brentwood, the senator from Illinois praised Schwarzenegger for passing a policy that could cut greenhouse gases associated with car and truck fuel 20 percent by 2020. "We know that transportation fuels account for a third of America's global warming pollution," Obama said at the press event. "And we know there are fuels available that emit less carbon dioxide into the atmosphere."

If Obama became president and adopted such a policy, it would potentially strike a major blow to the oil sands. A report from the nonprofit organization Ceres and the financial intelligence group RiskMetrics modeled what might happen. As states attempted to clean up their fuel supplies, one of their easiest options would be to ban fuel derived from northern Alberta. This could cut the market for oil sands by one-third, the report estimated. That wasn't even the worst-case scenario. "In the unlikely event that no options were available for Canadian oil sands producers to comply with the [low-carbon fuel standard]," the report read, referring to potentially billions of dollars of investments in technologies to lower the industry's climate footprint, "the U.S. transportation market could conceivably disappear."

That's why the industry's defenders in Canada became extremely sensitive toward anything uttered about the oil sands in the United States. When an organization representing American mayors adopted a resolution in 2008 supporting "federal legislation that prohibits government use of unconventional or synthetic fuels" derived from the oil sands, it barely made news in the U.S. But the decision got wall-to-wall coverage in Canada. "There, newspapers and other media outlets are crammed with articles and editorials denouncing the decision," a post on the site *Autoblog* explained.

As it looked more and more likely that Obama was going to win the election, the Harper government put together a defensive strategy. Should Obama become president, Canada would propose that the oil sands be exempt from climate legislation. In exchange, Harper would promise a secure supply of oil from a trusted ally instead of from producers in the Middle East. On November 8, Obama beat McCain by about nine million votes. In his victory speech, Obama said that a "defining moment of change has come to America." But Prime Minister Harper was seemingly unmoved by the historic election of America's first Black president. The morning following Obama's victory, Harper's cabinet ministers began publicly "calling for a pact [with the United States] that would keep emissions down while protecting Alberta's oil sands projects," the *Globe and Mail* reported.

Canada had often been seen as a friendly liberal neighbor to the north. But Harper's petro-realpolitik raised some eyebrows in Washington. "Seriously, Canada, just a couple of days into his transition, and already you're trying to play our Prez-elect," wrote Joseph Romm, founder of the progressive think tank the Center for American Progress. "Back off dudes!"

"A full-on barney"

TENSIONS WERE running a bit high at the Pebble Beach resort. This wasn't entirely surprising. Executives and public figures from Rupert Murdoch's media empire had flown in from across the world for a five-day event called "Imagining the Future." Holed up in a five-star California hotel situated between Monterey and Carmel, representatives from News Corporation's global array of conservative newspapers and television networks were trying to make sense of the latest edict from their boss: Murdoch wanted the media company to start taking climate change seriously. Whether through reducing greenhouse gas emissions from News Corp's internal operations or convincing its millions of readers and viewers to adopt more sustainable behaviors, going green was "a sound business strategy," Murdoch argued.

The 2006 gathering took place in an era of rising concern about the climate emergency. That was reflected in a presentation at Pebble Beach by Al Gore, in which the presidential candidate turned environmentalist did a screening of his new film *An Inconvenient Truth*. In a Q&A session afterward, Andrew Bolt, a columnist for the *Herald Sun* paper in Australia who frequently questioned the scientific consensus on climate change, got up to speak. According to an account in the *Bulletin*:

> Bolt opened his comments by congratulating Gore on his performance, then began to attack claims made by Gore in his film. Soon, according to one onlooker, the pair were involved in "a full-on barney." Gore ended up shouting at Bolt. "It was brilliant," says one onlooker. "Embarrassing," recalls another.

Bolt remembers the exchange differently. "I most certainly did not congratulate him on feeding the audience fake news," he later recalled. "Nor was there a 'full-on barney.' I calmly and politely asked him to explain three misrepresentations in his presentation, and he then went red and started shouting. This is more properly described as a meltdown."

But if outbursts from skeptics like Bolt made Murdoch reconsider his green ambitions, he didn't show it. The following year, he held an event in Manhattan that was broadcast to employees of Fox News, the *Wall Street Journal*, the *New York Post*, the *Australian* and dozens of other media outlets owned by News Corp. Murdoch said that the company would be launching a Global Energy Initiative, with the goal of becoming carbon-neutral by 2010. That was just the start. "Our audience's carbon footprint is ten thousand times bigger than ours. That's the carbon footprint we want to conquer," he said. The company's newspapers, TV stations, and movie studios would "revolutionize" how people saw climate change. "For too long, the threats of climate change have been presented as doom and gloom—because the consequences are so serious," he said. Murdoch wanted to make the issue "dramatic, make it vivid, even sometimes make it fun. We want to inspire people to change their behavior."

A decade later, it would seem unthinkable that Murdoch would give such a speech. But in the mid-to-late 2000s, public opinion on climate change was shifting rapidly. That could have been due to Gore's documentary, which won an Academy Award in 2006; a warning about the heavy financial impacts of global warming from former World Bank economist Nicholas Stern; dystopian disaster films such as *The Day After Tomorrow*; or increasing certainty from the Intergovernmental Panel on Climate Change, which warned that climate change is "very likely" being caused by humans. Whatever the reasons, the numbers of people worldwide who saw global temperature rise as a "very serious problem" grew significantly from 2000 to 2006, according to surveys from GlobeScan. In the United States nearly 50 percent of respondents now agreed with that statement. In parts of Europe, it was above 75 percent.

And elites were shifting their stances too. Despite the best efforts of the Global Climate Coalition and other spreaders of climate disinformation, there was a growing consensus among executives and politicians that the crisis demanded action—or at least the appearance of it. The CEO of Exxon, Rex Tillerson, seemed to acknowledge that new reality during a 2007 speech at the industry event CERAWeek, where he stated that "the risks to society and ecosystems from climate change could prove to be significant." Worrying about global heating was a sufficiently mainstream position that John McCain felt comfortable making it part of his presidential campaign. Even as he called for more offshore oil drilling, the Republican senator stated that "we need to deal with the central facts of rising temperatures, rising waters, and all the endless troubles that global warming will bring."

Murdoch was apparently convinced of the seriousness of climate change by his son James. "I don't think it happens without him," said David Folkenflik, NPR media correspondent and author of *Murdoch's World: The Last of the Old Media Empires*. Around this period, Murdoch's other son, Lachlan, was distancing himself from News Corp due to personal feuds with company executives. James at the time was chief executive of British Sky Broadcasting. The absence of Lachlan gave James, a political moderate who was becoming more interested in climate change issues through his wife, Kathryn, greater influence over his father, Folkenflik said. Whether it was due to family dynamics or a shifting elite discourse, Murdoch made his new perspective clear during a 2006 speech in Tokyo. "Until recently, I was somewhat wary of the warming debate," Murdoch said. But he had become convinced that "the planet deserves the benefit of the doubt."

Some of News Corp's papers started to echo their boss. A London-based paper called the *Sun* had previously run stories skeptical that climate change exists. But in 2006 it published a weeklong series on eco-friendly living and urged its readers to "Go green." That message was also promoted by the *New York Post* in its coverage of New York's Go Green Expo. And in Australia, the *Daily Telegraph* started running editorials urging the federal government to take climate change more seriously. But as Bolt's fight with Gore at Pebble Beach showed, some people within News Corp's media empire remained skeptical. A former environmental reporter at the *Australian* would later say it was "torture" to get her editors' support for climate coverage, an accusation the paper denied afterward.

Even still, Murdoch was convinced he could get some of News Corp's most conservative voices interested in covering the issue productively. During a rare 2007 interview with the environmental outlet *Grist,* Murdoch said he was planning to have conversations about it with Fox News anchors Sean Hannity and Bill O'Reilly. "Probably Sean's first reaction will be that this is some liberal cause or something, you know? But he's a very reasonable, very intelligent man," Murdoch argued. "He'll see, he'll understand it. As will Bill—he just likes to get debate going between people. And that has its benefits—someone says, 'No there isn't,' someone says, 'Yes there is,' and they have it out for ten minutes and it's entertaining and creates more consciousness."

The head of Fox News, Roger Ailes, was much less enthusiastic than Murdoch. Around this time, Fox Studios created a public service announcement featuring Ailes and Kiefer Sutherland, star of the studio's hit TV series *24*. "Global warming is a crime for which we are all guilty," Sutherland said in the video. Ailes's contribution was comparatively muted. "I was very clear," he says in the video, "that energy was gonna be one of the things that was going to determine leadership for countries in the future." To Folkenflik, it sounded like Ailes "came up with something that sounded supportive but committed him to nothing."

Nevertheless, by the end of 2008, the United States was in a moment of extraordinary possibility for stopping climate change. Barack Obama was president and Democrats controlled both chambers of Congress, meaning they had the political power to pass legislation that could start to phase out oil, coal, and gas from the economy. Prominent Republicans like McCain supported efforts that could achieve that goal, including making polluters pay a price for their carbon emissions. Murdoch, owner of media outlets with vast power to swing public opinion, seemed eager to help fix the crisis. But within a year and a half, that window of opportunity would slam shut. And the oil sands industry, along with the Canadian government, deserves a significant portion of the blame.

"Public embarrassment"

TO OUTSIDERS who didn't pay that much attention to Canada, it might have seemed as though the government was taking climate change extremely seriously. During a 2007 speech in Berlin, Germany, prime minister Stephen Harper called global warming "perhaps the biggest threat to confront the future of humanity today." Canada may only produce 2 percent of global emissions, he said, "but we owe it to future generations to do whatever we can to address this world problem. Canadians, blessed as we are, should make a substantial contribution to confronting this challenge." Harper told his international audience that Canada was investing heavily in greener forms of energy. "Just as the Stone Age did not end because the world ran out of stones, the Carbon Age will not end because the world runs out of fossil fuels," he said.

Just five years earlier, Harper had sent out a fundraising letter deeming the Kyoto Protocol a "socialist scheme" that would "cripple the oil and gas industry," while referring to climate science as "tentative and contradictory" and calling carbon dioxide a gas "essential to life." Had the former Imperial Oil employee truly changed his views? The network of industry-backed climate deniers that had celebrated Harper becoming prime minister didn't think so. "I don't believe the government necessarily believe what they're saying at all," claimed Tom Harris, who had helped organize the Imperial Oil event in Ottawa where Patrick Michaels and dozens of others attacked the science behind Kyoto.

Harris was correct. Though Harper talked about the urgency of climate change on the world stage, in Ottawa the Conservative leader did everything he could to suppress the science creating that urgency. Several months after Harper's Germany speech, Environment Canada brought in a new media relations policy restricting the ability of scientists who worked at the department to speak to journalists. If a journalist were to get in touch seeking an interview, those scientists would immediately have to refer the request to a Harper-government-approved "Media Relations Officer." The officer might then dictate "approved lines" that the scientist could use in the interview, do the interview on behalf of the scientist, or delay it so long that the journalist eventually gave up.

The implications of this new policy went far beyond Canada's borders, given that the country's scientists were producing globally relevant work. More than two hundred Canadian scientists had contributed to the Intergovernmental Panel on Climate Change's research, an organization that by 2007 was calling it "unequivocal" that humans are causing global warming. Many of these scientists were employed by the federal government. Harper soon went one step further than blocking their access to reporters. In addition to making it harder for interested journalists to interview federal scientists, he would bury the scientific work that created media interest in the first place.

In December 2007, scientists with Natural Resources Canada completed a 450-page report called *From Impacts to Adaptation: Canada in a Changing Climate*, which featured the work of 145 authors and 110 expert reviewers. The report documented "the advances made in understanding Canada's vulnerability to climate change during the past decade." Previous prime ministers might have held a press conference to release the report, but Harper's government refused for months to make the work of his own scientists public. Finally, in March 2008, the government quietly posted the report on its website. The report appeared late on a Friday, which was "the worst possible moment for attracting any further interest from the media," according to an organization called the Climate Action Network. "Unsurprisingly, the report's contents generated little media coverage, frustrating many of the contributing government scientists, public servants, and academics."

This pattern played out again later that year, as scientists with Health Canada prepared to a release a five-hundred-page report forecasting the illnesses, respiratory and cardiovascular disorders, and premature deaths that would likely be caused in Canada by warmer-than-average temperatures—health impacts that could fall especially hard on Indigenous peoples due to their relative lack of access to high-quality care. The scientists behind the report had been told that it would be pitched to Canadian media and include events across the country. But on July 3, 2008, the scientists were given an update: it would be a "low-profile release" instead. The report came out on a Thursday afternoon in the middle of July.

Then there was the incident involving Don MacIver, a government climate researcher who was scheduled to present his work about climate impacts and adaptation opportunities to an international conference in Poznan, Poland, on December 5, 2008, only to find out his trip had been canceled by Environment Canada—literally on his way to the airport. The scuttling of his travel was apparently approved by Conservative environment minister Jim Prentice, who, after reviewing MacIver's presentation slides, determined the scientist's contribution to the conference wasn't necessary. MacIver, calling the entire thing a "public embarrassment to Canada," resigned from the government shortly after.

These events occurred at the same time Harper was appointing climate change deniers to key scientific positions. One of the first was Christopher Essex, a professor of applied mathematics at the University of Western Ontario, who was appointed to the board of the Natural Sciences and Engineering Research Council of Canada in November 2006. Essex had previously gone on record arguing that "the claim that there is a global warming crisis threatening to bring chaos and destruction upon the world is so feeble you were probably feeling skeptical anyway. You were right." He later contributed to a study claiming "there is no compelling evidence that dangerous or unprecedented changes are underway" in the Earth's climate.

Harper also appointed to the research council board Mark Mullins, then executive director the Fraser Institute, which is one of the country's most conservative think tanks. The institute had by then been attacking climate science for over a decade, publishing studies with titles such as "Greenhouse Gas Reductions: Not Warranted, Not Beneficial," and "The Science Isn't Settled." Tax filings show that in the early 2000s, Exxon, Imperial Oil's parent company, gave $120,000 to the Fraser Institute for work related to "climate change." Later that decade, the think tank received $325,000 from charitable foundations linked to Koch Industries. "The Institute publicized a wide variety of Canadian and U.S. government data on the environment, demonstrating the exaggerations of environmental alarmists," Mullins bragged in a report celebrating the think tank's thirty-fifth anniversary. The report's introduction was written by Prime Minister Harper himself, who praised Mullins's "commitment to the highest standards of research."

The same day that Harper gave a top scientific position to Mullins, he also appointed John Weissenberger to the board of the Canada Foundation for Innovation, which funds research into climate change and other issues. Weissenberger was a personal friend of Harper's and a geology professor at the University of Calgary who had done consulting work for Husky Energy and other oil and gas companies for more than two decades. "To those who doubt the scientific basis of global warming theory, we say: don't let a cabal of government-funded scientists, environmental activists and journalists convince us they're the mainstream," Weissenberger said in 2006.

Harper was in a sense employing the same tactics that executives at his former employer Imperial Oil had used during the 1990s and early 2000s—but now on a much more consequential scale. His government's own scientists were producing world-class data and projections about the climate emergency. Yet rather than acting on that life-saving information by shifting a national economy based heavily on oil and gas into greener industries, Harper prevented this science from reaching the public and used his position as prime minister to legitimize and give a huge platform to climate deniers. It was the same thing Imperial Oil had done with the inconvenient climate research produced by its own scientists.

None of this addressed the growing crisis south of the border, however, where Obama and his political allies were moving forward with climate legislation that could deal a fatal blow to the oil sands. Starting in the late 2000s, Harper's government backed a secret effort to counter that threat, an effort that would mimic the dirty tricks of oil companies while destroying a key part of the Obama administration's environmental agenda.

"They're struggling forward"

DECEMBER 30, 2009, was a bad day for an American named Michael Whatley. Based in Washington, D.C., on K Street, an area adjacent to Capitol Hill infamous for its concentration of lobbyists and corporate operatives, Whatley was founding partner of a consulting firm called HBW Resources, which counted among its clients major oil and gas companies, many with large investments in the oil sands. Whatley's day was not going well because eleven governors of Northeast and Mid-Atlantic states had just agreed to consider adopting a climate change policy that would limit their states' "reliance on petroleum-based fuels." That was bad news indeed, Whatley thought. He fired off an email to his contact in the Canadian embassy: "Please let me know your thoughts."

Whatley came to this climate change fight with a background in conservative politics and religion. The D.C. lobbyist held a "juris doctorate and master's degree in theology from the University of Notre Dame," according to his bio, as well as a "master's degree in religion from Wake Forest University." He later served as attorney and senior policy adviser on George W. Bush's first presidential campaign and transition team, and was then appointed chief of staff to senator Elizabeth Dole, a former cabinet secretary and the wife of Republican elder statesman Bob Dole. After his stint in government, Whatley began working at HBW, which had been founded in 2005 to "provide counsel and guidance to companies regarding governmental and industry energy initiatives, including issues related to energy security, exploration and production, advanced technologies, air quality and refining."

Here's how that looked in practice: Sometime in the late 2000s, Whatley's firm HBW created the Consumer Energy Alliance, an organization devised to have the look and feel of a "grassroots" social movement. It claimed on its website to be "a voice for consumers interested in vital public issues" when in reality it was supported by BP, Chevron, Exxon, Marathon, Shell, and other oil producers with financial ties to the oil sands. To environmental groups, the guise wasn't all that hard to see through. "They're a front group that represents the interests of the oil industry," Luke Tonachel, an analyst with the D.C.-based Natural Resources Defense Council, said at the time.

The Canadian to whom Whatley sent off his email in late 2009 was Gary Mar, a diplomat from Alberta based in the modernist Canadian embassy on Pennsylvania Avenue. It was unusual for Canadian provinces to have full-time lobbyists in the American capital. But with billions of dollars of U.S. investment flowing into the oil sands, and millions of barrels of bitumen in turn flowing to refineries south of the border, it made sense for Alberta to have a political presence in Washington backed by the diplomatic resources of the Canadian government. Mar himself had the extroversion of an American. He was known to start belting out Elvis songs in the middle of meetings and once posed ostentatiously for a newspaper photo in front of portraits of Winston Churchill and John F. Kennedy.

It wasn't long before Mar had infiltrated Capitol Hill's fossil fuel elite. "It's amazing he's not cloned somehow," said Cindy Schild, a manager with the American Petroleum Institute. Mar, she explained, "is everywhere. He knows everybody." Whatley himself referred to him as "a force of nature." Mar's presence wasn't only confined to Washington. Anytime that state policymakers tried to bring in climate change legislation reducing the amount of oil sands fuel consumed on their roads, Mar would leave the capital, ready to charm and cajole. "I found myself spending a great deal of time trying to influence state governments," he would later claim. "I have had influence in stopping legislation that would have been unfairly harmful to Alberta's interests in Minnesota, Michigan, and Maryland."

By the late 2000s, "harmful" climate legislation seemed to be everywhere. One of the biggest threats was the low-carbon fuel standard, the policy that had been first adopted by California in 2007. So when Whatley heard in December 2009 that the eleven East Coast governors were considering a low-carbon fuel standard in order to alleviate what they deemed the "serious risks" of climate change, he knew that he'd have to act fast. Writing to Mar, Whatley explained that the governors' plan warrants "a very serious response on all levels." Whatley and Mar's exchange was contained in hundreds of pages of personal emails that were made public due to a freedom of information request to the Alberta government. "As we have discussed, this fight cannot take place within DC—and we need to get a team funded and on the ground in these 11 states as soon as possible," Whatley wrote.

But the battle strategy quickly needed to grow larger. In early 2010, an additional ten Midwestern states announced they were also studying a low-carbon fuel standard. Whatley sent an official proposal "on behalf of HBW Resources" to Mar in late January. The lobbyist said that with the support of government officials from Canada, he would "defeat efforts" to develop low-carbon fuel standards in "Northeast, Mid-Atlantic and Midwestern states." He pledged as well to "address potential efforts" to adopt cleaner fuels legislation in "Washington, Oregon, Michigan, Minnesota and other states." How would Whatley accomplish all this? One option that he recommended was "conducting a grassroots operation" in "target states" that would "generate significant opposition to discriminatory low carbon fuels standards."

Whatley already had a track record of success. During Barack Obama's first year in office, the president had made it clear that he viewed the heavy environmental impacts of Canada's bitumen as a serious "dilemma," saying that "what we know is that oil sands creates a big carbon footprint." By the following year, Democratic Congress members were attempting to make good on that promise by passing legislation that would put a cap on U.S. greenhouse gas emissions and allow polluters to purchase and trade emissions credits if they couldn't meet the cap in time. (This was the plan created by environmental policy experts decades earlier to deal with the problem of acid rain). Alongside this cap-and-trade bill, some Congress members were proposing climate legislation that Obama himself had campaigned on: a national low-carbon fuel standard.

Environmental groups saw the policy as a big deal for slowing down climate change. If passed federally, a low-carbon fuel standard could signal "the end of the petroleum age," argued the Natural Resources Defense Council.

In August 2009, Whatley's Consumer Energy Alliance ran radio and TV ads in Tennessee, Montana, and the Dakotas, warning that low-carbon fuel standards "threaten thousands of American jobs" and "would be disastrous for American consumers." Each ad told viewers to complain to their state's representatives in Congress and provided phone numbers to make it easier. One of those numbers belonged to Tennessee senator Lamar Alexander, a Republican who had once observed that a national low-carbon fuel standard "makes a lot of sense." As the ads ran in Alexander's home state, Whatley pressured the senator in Washington, D.C. "I am working a deal to keep Lamar Alexander from offering a [low-carbon fuel standard] amendment," Whatley wrote to Mar on September 30, 2009. "If we can keep him off of it—it will die an ugly partisan death on the Senate floor." Soon after, the senator told Knoxville media he was undecided on the policy he had once favored.

While this was happening, a low-carbon fuel standard was also being proposed in the House of Representatives. That attempt ran into fierce opposition from an organization called the Center for North American Energy Security, which represented Chevron, Exxon, and other producers active in the oil sands. First, the group identified policymakers and congressional staffers who were potentially unsympathetic to the fuel standard. "Then we talked to those people to a) alert them it exists, b) explain why it was a mistake, and c) try to get support to repeal it," said the center's leader, Tom Corcoran, himself a former Republican congressperson from Illinois.

With political opposition mounting, Democrats realized that the bigger cap-and-trade legislation they were drafting wouldn't make it past the committee stage with a fuel standard attached. "So they deleted the low-carbon fuel standard," Corcoran explained. The Canadian government was "aware of what we were doing and supportive of it," he claimed.

With the national threats taken care of for the time being, Mar and Whatley focused on state-level standards. Their strategy was laid out in Whatley's 2010 campaign proposal. First off, it explained, they needed to team up with "affiliated energy coalitions and trade associations, thought leaders, elected officials, unions and key allies." The goal was to enlist these players to "build opposition" against low-carbon fuel standards "in each of our target regions." The campaign also needed "state-based and regional 3rd party advocates for Canadian oil sands" to give it legitimacy. So item number one on Whatley's "Action Plan" was to develop "easy-to-read and user friendly informational briefs" for airlines, truckers, railroads, shippers, and other "energy consumer groups."

With the proper motivation, Whatley's proposal explained, these groups could "generate op-eds and letters to the editor of regional and local newspapers." And they could also "write letters to governors and key elected officials."

This supposed popular groundswell would then be legitimized further, it explained, by a select group of "thought leaders," those public intellectuals with the ear of political power. Whatley's proposal suggested engaging with seven think tanks, two of which, the Cato Institute and the Heritage Foundation, were big players in the climate change denial movement, having received millions of dollars in funding over the years from Exxon and Koch Industries. The Consumer Energy Alliance would meanwhile run anti-fuel-standard ads, coordinate with "key allies" such as the American Petroleum Institute, lobby policymakers, and generate media attention.

If everything went to plan, Whatley's briefing concluded, "HBW Resources will be able to successfully draw critical local, state and regional attention to the adverse impacts of efforts to restrict imports of Canadian oil sands into the United States."

One of the campaign's first victories came in mid-April of 2010, when Wisconsin abandoned its low-carbon fuel standard. Unable to visit public hearings in the state capital, Madison, because of a snowstorm, Mar had gotten two Canadian diplomats to read a statement opposing the policy. That intervention infuriated local scientist Peter Taglia, who later said in an interview that he "was disappointed with the Canadians... They behave basically the same way the Texas oil companies do." But the Consumer Energy Alliance was ecstatic about Wisconsin's decision. "The removal of the economy-killing [fuel standard] is good news for consumers in the Badger State," read a statement on its website.

That June, Alberta environment minister Rob Renner joined the fight. Officially, he was making a visit to the U.S. Northeast as part of a "Clean Energy Mission." But in reality, Renner was working with Mar and Whatley to create opposition to the low-carbon fuel standard that eleven governors were still considering. The Consumer Energy Alliance sponsored Renner for an anti-fuel-standard forum in Boston, which generated media stories in eighteen outlets. Other groups at the event included the Massachusetts Motor Transportation Association and the Associated Industries of Massachusetts. "We have been assured by several of the participants in the forum that they will be willing to send letters to their governors, the federal Congress and the Obama administration opposing a discriminatory LCFS," Whatley reported to Mar, using the shorthand for low-carbon fuel standard. The Albertan responded ten days later: "Thanks for keeping me several steps ahead of other advisors." Whatley replied, "Thanks for being great to work with."

By the end of 2010, the Whatley-Mar plan had largely succeeded. A fake grassroots campaign led by an oil industry operative and a diplomat working out of the Canadian embassy had blunted a key piece of Obama's climate agenda, one that would for the first time have imposed heavy carbon regulations on the oil sands and other pollution-heavy sources of road fuel. Even though such legislation was still being considered by several states, the environmental zeitgeist behind it was clearly weakened. By then, California was the only U.S. jurisdiction implementing the policy. National interest had collapsed. There was little support for the standard in the Midwest. The eleven Northeast and Mid-Atlantic states were moving forward without political conviction, allowing the standard to be analyzed, and occasionally talked about, but never actually implemented.

"Global warming!"

IT WAS EARLY 2010, and Barack Obama was attempting to rally Americans in support of economy-altering government action to stop climate change. "I know there have been questions about whether we can afford such changes in a tough economy," the president said in his annual State of the Union address, "and I know that there are those who disagree with the overwhelming scientific evidence on climate change." Obama paused for emphasis. "But even if you doubt the evidence, providing incentives for energy efficiency and clean energy are the right thing to do for our future. Because the nation that leads the clean-energy economy will be the nation that leads the global economy. And America must be that nation."

Few people were aware of it, but these talking points had been developed in part through a collaboration between Rupert Murdoch's company News Corp and a Republican operative and public opinion researcher named Frank Luntz, who was notorious for authoring a playbook years earlier instructing the GOP on how to undermine climate science. But times had changed. Murdoch was now attempting to get his media empire to play a more proactive role in addressing the climate emergency. And with the Obama administration backing a cap-and-trade climate bill that would force polluters to pay a price for emitting greenhouse gases into the atmosphere, News Corp reached out to Luntz and asked if he'd develop a strategy for getting a potentially skeptical U.S. public on board. "Luntz wanted to get on the right side of history and had a client willing to pay him for it, so he began figuring out how to sell the idea," Eric Pooley writes in *The Climate War.*

Luntz did polling, which found that far-reaching climate legislation could be hugely popular, even among deniers of the scientific consensus, if it were framed as a way to reduce U.S. dependence on Middle Eastern oil and prevent the country from losing out on clean-energy jobs to China. When cap-and-trade was described in those terms, Luntz argued, "Americans overwhelmingly support the policy." Pooley writes that "Luntz had been briefing senators, including Lindsey Graham, about this messaging strategy. And after Graham mentioned it to Rahm Emanuel at the White House, it turned up in Obama's State of the Union speech."

This was arguably one of the more useful things that Murdoch's media empire had ever done to fight climate change. But positive messaging on global warming could only take Obama so far. His ability to enact a policy that would help sever the link between burning fossil fuels and economic growth wouldn't depend on persuasion or clever framing; it was about raw political power. And in 2010, few companies had more of that than Koch Industries.

Koch Industries was the second-largest privately held company in the country at the time of Obama's inauguration, bringing in annual revenues of $100 billion. It was also one of the largest and most consequential players in Canada's oil sands industry. The Pine Bend Refinery in Minnesota refined roughly 25 percent of all oil sands crude entering the United States. "It was specifically designed to process heavy, sour crude piped in from Canada," a Koch-produced newsletter from 2010 explained. That reliance on an especially polluting fuel source—in addition to the company's ownership of coal power stations, petrochemical plants, and pipelines—left Koch Industries highly vulnerable to emissions-reducing legislation.

Beginning in the 1990s with the early Cato Institute conference featuring Patrick Michaels and other professional deniers, Koch Industries contributed heavily to groups that spread disinformation about global temperature rise. From 2005 to 2008, a Greenpeace report calculated, charitable foundations linked to the company gave $24.9 million to more than forty "climate denial and opposition organizations." "This is only part of the picture," the report noted, "because the full scope of direct contributions to organizations is not disclosed by individual Koch family members, executives or from the company itself."

But even if the funding sources were opaque, the impacts of all that spending were clearly visible. During Obama's first month in office, Americans began seeing a TV advertisement featuring a rich young man next to a plate of canapés. "Hey there," he told viewers, "I'm Carlton, the wealthy eco-hypocrite. I inherited my money and attended fancy schools. I own three homes and five cars, but always talk with my rich friends about saving the planet. And I want Congress to spend billions on programs in the name of global warming and green energy. Even if it causes massive unemployment, higher energy bills, and digs people like you even deeper into the recession. Who knows, maybe I'll even make money off of it!"

That $140,000 advertising campaign was paid for by Americans for Prosperity, a far-right political organization founded and funded by the Koch brothers. At the time, the organization was helping organize and coordinate a series of spontaneous-looking grassroots rallies across the United States in opposition to Obama. David Koch denied he had anything to do with this burgeoning "Tea Party" movement, but a speech he gave in October 2009 at the Crystal Gateway Marriott hotel in Arlington, Virginia, suggested otherwise. "Five years ago, my brother Charles and I provided the funds to start Americans for Prosperity, and it's beyond my wildest dreams how

AFP has grown into this enormous organization," Koch said. "We envisioned a mass movement, a state-based one, but national in scope, of hundreds of thousands of American citizens from all walks of life standing up and fighting for the economic freedoms that made our nation the most prosperous society in history."

An unnamed Republican insider quoted by Jane Mayer in the *New Yorker* made the connection explicit. "The Koch brothers gave the money that founded [the Tea Party]," he said. "It's like they put the seeds in the ground. Then the rainstorm comes, and the frogs come out of the mud—and they're our candidates."

By the time the 2010 midterm elections rolled around, there were 129 Tea Party–supported candidates running for the House and nine for the Senate, all Republicans. One out of three won their races as the Republicans took back the House, dealing a major blow to Obama's agenda. The victors included Minnesota representative Michele Bachmann, a noted purveyor of far-fetched conspiracy theories, who while campaigning said, "I want people in Minnesota armed and dangerous" in the fight against a cap-and-trade approach to lowering emissions because "the science indicates that human activity is not the cause of all this global warming... in fact, nature is the cause, with solar flares, etc."

Tea Party protests were in their early days animated by populist rage against Obama's bank bailouts and health care plan, as well as not-so-subtle racism toward America's first Black president. One widely circulated Tea Party poster portrayed Obama as an African witch doctor. But around the time of the "Carlton" advertisement, Americans for Prosperity started urging protesters to turn their anger against the cap-and-trade legislation making its way through Congress. The group spent more than $5 million taking a seventy-foot-tall hot-air balloon around the country that displayed a slogan opposing the climate change bill: "Higher taxes, lost jobs, less freedom."

By February 2010, making sure cap-and-trade legislation didn't pass was one of the top goals for Tea Party activists. When a Houston attorney named Ryan Hecker proposed that activists vote online for their top ten priorities, "reject cap & trade" was selected as number two. The Obama administration believed a key selling point of the legislation was that its market-based approach of carbon trading would appeal to conservatives. Indeed, one of the bill's cosponsors was Republican senator Lindsey Graham. But as Tea Party protests picked up in his home state of South Carolina, Graham began to waver in his support, warning his Democratic colleagues, "It's gonna become just a disaster for me on the airwaves." On April 24, Graham withdrew. Several months later, with Republicans and some conservative Democrats refusing to vote for the legislation, cap-and-trade died.

The collapse of a bill that had proposed to cut U.S. greenhouse gas emissions to 17 percent below 2005 levels by 2020 was one of the biggest climate policy failures in history. But the Koch-backed Tea Party disinformation campaign on cap-and-trade had implications beyond just legislation. Three years earlier, Rupert Murdoch had stated his belief that he could convince Fox News hosts like Sean Hannity to cover the climate emergency more seriously. Instead, Hannity, Glenn Beck, Neil Cavuto, and other network anchors became some of the biggest media cheerleaders of the Tea Party, in some cases even broadcasting live from Tea Party events and encouraging viewers to join up with local chapters.

It was always an open question whether Fox News would follow its boss Murdoch's suggestion that it report in good faith on climate change. But in late 2010, Fox News VP of News Bill Sammon—who had earlier that year stated his support for the Tea Party, claiming that mainstream media "hates" the movement—slammed the door shut entirely. "Given the controversy over the veracity of climate change data," Sammon wrote in an email to Fox News staffers, which was later leaked, "we should refrain from asserting that the planet has warmed (or cooled) in any given period without IMMEDIATELY pointing out that such theories are based upon data that critics have called into question. It is not our place as journalists to assert such notions as facts, especially as this debate intensifies."

The political shift was evident in other ways. While campaigning for president in the 2008 election, Republican candidate John McCain had warned that "unless we reverse what is happening on this planet, my dear friends, we are going to hand our children a planet that is badly damaged." An article on the Fox News site helpfully informed viewers at the time that "McCain favors climate-change legislation that would set caps on carbon and other greenhouse gas emissions and offer incentives for industries to come up with new energy sources."

But by 2010, Fox News was referring to "man's alleged impact on the environment." That same year, South Carolina Republican representative Bob Inglis was beaten in the primary by a Tea Party candidate after saying on a local radio show that he believed climate change is real. In November of 2011 Newt Gingrich called the TV ad he'd made with Nancy Pelosi urging action to lower emissions "probably the dumbest single thing I've done in recent years," His new tack: "I actually don't know whether global warming is occurring."

Murdoch's public views on climate change also appeared to shift. "Climate change has been going on as long as the planet is here, and there will always be a little bit of it. At the moment, the North Pole is melting, but the South Pole is getting bigger," Murdoch said in 2014. Not long after, the owner of News Corp tweeted an aerial photo of a frozen Arctic landscape. "Just flying over N Atlantic 300 miles of ice," he wrote. "Global warming!"

Denial was once again resurgent in U.S. media and politics, but the situation looked different in the courts, where a legal effort to expose the lies of oil and gas companies was just getting started.

VI

The Climate Goes to Court

(2008–2014)

“The island is sad that it’s going away”

IT HAD BEEN A DECADE since Steve Berman helped sue Big Tobacco and force the industry to pay a $206 billion settlement. That victory alone was a career-making triumph. But in the years that followed, Berman set his sights on an even bigger adversary, one with so much financial power and political might that it made the cigarette companies seem tiny in comparison. The Seattle attorney wanted to sue the oil and gas industry. He was convinced it had lied to the public about climate change the same way Big Tobacco had lied about cancer. “I’ve always been interested in environmental issues,” he explained. Now he just had to find the right case.

In the late 2000s, Berman heard about a coastal village in Alaska called Kivalina whose Indigenous inhabitants would have to be relocated because of rising oceans. The Army Corps of Engineers had done a study about the situation and concluded that climate change was to blame. It estimated that the cost of relocation could be up to $400 million. Few people had brought forward major climate change lawsuits at this point, owing in part to the difficulties of attributing a complex global phenomenon with many causes to any individual company or industry, but Berman saw potential. Not only had an arm of the federal government directly blamed global temperature rise for Kivalina’s predicament, it had assigned a specific cost estimate to the damages. “I go, ‘wow, what an awesome platform to begin global warming litigation,’” Berman said.

The conventional thinking at the time was that climate change is a problem for which we are all partially responsible. Since our society is built on gasoline-burning vehicles, coal-burning power plants, and all manner of polluting industries essential to our way of life, you couldn't place blame on a particular company or industry. But Berman wasn't so sure. He'd heard a similar claim used a decade earlier by tobacco companies: that threats to people's health were everywhere, so you couldn't single out cigarette makers. What caused that argument to collapse in court was documents showing that tobacco companies had known internally about the dangers of their products but tried to convince the public they were safe. This, Berman suspected, was exactly what fossil fuel companies were doing with climate change.

Berman met with attorneys who'd previously represented the village, along with a few other lawyers interested in the case. They agreed that it had potential for success. In 2008, Berman's legal team filed a lawsuit on behalf of Kivalina against Exxon, Chevron, ConocoPhillips, BP, Shell, and more than a dozen other fossil fuel producers. The legal complaint noted that these companies are a major source of greenhouse gas emissions through their vast global operations, which included their investments in the oil sands. But more importantly, the complaint accused those companies of deliberately spreading doubt and uncertainty about the science of climate change. "Relying on tactics developed by the tobacco industry to discredit health risks associated with tobacco use, ExxonMobil has channeled $16 million over the 1998 to 2005 period to 42 organizations that promote disinformation on global warming," it read.

Facts like this were the key to Berman's case. It wasn't so much that Exxon and its competitors had sold environmentally damaging products. There was currently no law against that. It was that the fossil fuel industry had intentionally deceived the public in order to protect its profits. That was fraud. There were clear parallels to the earlier lawsuit against Big Tobacco. "You're not asking the court to evaluate the reasonableness of the conduct... You're asking a court to evaluate if somebody conspired to lie," Berman told the *Atlantic* at the time.

The villagers living in Kivalina were not there entirely by choice. For thousands of years, the barrier reef island in northwest Alaska had been used as a summer hunting location for the Kivalliñigmiut Iñupiat tribe, whose members moved to the mainland in the winter to follow caribou migrations. But in the 1920s, the U.S. government built a school on the island and forced Iñupiat peoples to attend as part of a racist colonial effort to "civilize" Indigenous children. The seasonal hunting settlement in Alaska became permanent, and by the late 2000s there were about 374 people living next to the Chukchi Sea.

The town had traditionally been protected from the ocean by thick sea ice, which surrounded the coastline. But as global temperatures increased, the ice formed later and melted earlier, resulting in the shoreline eroding. This, combined with more frequent sea storms that are also linked to climate change, posed an existential threat to the entire community. In less than a generation, the Army Corps concluded, the island could disappear. Berman flew to Kivalina at one point early in the lawsuit and saw waves lapping ever closer to the school. Despair was pervasive. "The island is sad, sad that it's going away," he said.

A Kivalina Elder named Lucy Adams told one visitor that the town was becoming unrecognizable to her. "Thirty years ago, we built sod igloos. Men would get wood with dog teams. The dogs would pull the boat out onto the ice. Women would run alongside the dogs. I would run for hours and hours and never get tired. Everything was so clean. Clean water, clean fish. We burned trash. There were no plastic bags. We washed clothes by hand. If we ran out of wax candles, we used seal blubber. Everybody worked and exercised every day," she said. Adams sighed and looked up at the ceiling as she contemplated what had happened since. "We are standing alone to try and survive," she said.

Berman wasn't the only lawyer representing the village. Joining him on the lawsuit was Matt Pawa, an environmental attorney, and also Steve Susman. Berman had history with Susman. During the tobacco trials, they had actually faced off in court: Berman representing state attorneys general and Susman defending Philip Morris. But in the intervening years Susman, like Berman, had grown interested in climate change litigation. Susman even taught a course on it at one point at the University of Houston Law Center. In their legal world it wasn't uncommon for a lawyer such as Susman, who worked with both defendants and plaintiffs, to be representing a corporation on one case and the victims of malpractice in the next. "So we wound up working together," Berman said. "I've known Steve for a long time."

Their complaint accused prominent oil and gas companies of engaging in a "civil conspiracy" to "mislead the public about the science of global warming." It named "notorious climate skeptics" like Patrick Michaels as the public face of denial efforts. "The tactics employed in this campaign include the funding and use of 'global warming skeptics,' i.e. professional scientific 'experts' (many of whom are not atmospheric scientists) who regularly publish their marginal views expressing doubts about numerous aspects of climate change science in places like the *Wall Street Journal* editorial page but rarely, if ever, in peer-reviewed scientific journals," the complaint read.

The conspiracy section focused heavily on the Global Climate Coalition, the "outspoken and confrontational" fossil fuel industry group active during the 1990s and 2000s, which played a key role in convincing the United States not to ratify the Kyoto Protocol. Berman believed that a 1995 document provided evidence that coalition members were deliberately "distorting the public debate." It was the internal primer in which a coalition scientist explained that the contrarian arguments of Michaels "are not convincing." And in a document from the following year, coalition members acknowledged that climate change could cause "potentially irreversible" impacts including "significant loss of life." "But for years afterward," the complaint read, "the GCC and its members continued to tout the contrarian theories about global warming."

The impact of these actions was to preserve the profits of fossil fuel companies, the complaint alleged, by delaying policies that could lower greenhouse gas emissions. And this was how it all related back to Kivalina: absent global action to stop the climate emergency, the ocean was swallowing an Indigenous community, causing potentially $400 million in damages. It was a novel legal argument, given force by Berman's impressive courtroom track record. "But the hurdles may be greater than those in the tobacco wars," the *New York Times* reported. "Global warming is a diffuse worldwide phenomenon; a successful public nuisance case requires that defendants' behavior be directly linked to the harm." It was true—Berman's argument was largely circumstantial. To win in court, he would need to provide a more explicit link between the climate pollution of oil companies and the waves lapping up against Kivalina. But before that, Berman's team had to convince a judge the case was even worth considering.

In October 2009, a U.S. district court ruled against the Kivalina lawsuit. Berman appealed, but the case was again dismissed in 2012. Both times, federal judges decided that the question of who is to blame for climate change is best left to entities like Congress or the White House, which have the power to create environmental legislation. The judicial system, the judges both argued, shouldn't interfere with that legislative power. But the real legal issue was causation. How could you prove that oil dug out of the ground by Exxon was causing a tiny Alaskan village to disappear? "There is no realistic possibility of tracing any particular alleged effect of global warming to any particular emissions by any specific person, entity, group at any particular point in time," wrote U.S. district judge Saundra Brown Armstrong.

One of the first major global warming lawsuits was a failure. But Berman knew well the history of smoking litigation: it had taken dozens of lawsuits over decades before Big Tobacco's defenses had crumbled. And the ruling of Armstrong and other judges had left open several legal paths forward. A case like Kivalina likely couldn't proceed in federal court, but the dismissal made no mention of cases filed in state court, where the legal conditions were more favorable. And, crucially, the judges that dismissed Berman's complaint never offered an opinion on the conspiracy allegations. The question of whether the oil and gas industry had lied in order to delay action on climate change was still, as far as the courts were concerned, unsettled.

"A way to justify exploitation"

TO ANYONE SCHOOLED in the oil and gas industry's history of spreading climate disinformation, Steve Berman's Kivalina lawsuit might have read like a "who's who" guide to the most influential players. The complaint listed contrarians such as Patrick Michaels; polluter associations like the Edison Electric Institute and the Western Fuels Association; and industry-backed denial groups such as the Global Climate Coalition, the Advancement of Sound Science Coalition, the Greening Earth Society, and the Cooler Heads Coalition. But the lawyer's complaint also named an "industry-sponsored front group" that might have been unfamiliar even to seasoned disinformation experts. That was the Fraser Institute, the far-right Canadian think tank based in a city that thought of itself as one of the "greenest" places in the world.

You would hardly guess while walking past the Fraser Institute's headquarters in Vancouver that you were passing a piece of libertarian history. The think tank is on the fourth floor of a nondescript office building above a car dealership. To the north are the gleaming energy-efficient glass skyscrapers of downtown, built to LEED-certified sustainability standards, and beyond them the snowcapped mountains of Vancouver's North Shore. Within walking distance of the Fraser Institute is the headquarters of the David Suzuki Foundation, a prominent environmental group named after a Canadian ecological icon. And a short drive west is the building in the neighborhood of Kitsilano where Greenpeace opened its first office in 1971.

The Fraser Institute was founded in 1974, just seven years after Sun Oil's Great Canadian Oil Sands project began pumping its first bitumen. Like former Sun Oil president J. Howard Pew, the Fraser Institute was heavily influenced by the anti-communist ideology of the economist and conservative activist Friedrich Hayek. The institute's founder, an industrial chicken factory owner in the U.K. named Antony Fisher, was in fact good friends with Hayek. It was Hayek who apparently convinced Fisher to use profits from his chicken factories to create a free-market think tank in London called the Institute of Economic Affairs. Fisher later lost much of his fortune investing in a turtle farm on the Cayman Islands, a financial blow that he blamed on environmental and animal-rights activists. "Newly under-employed," as a short biography on the Institute of Economic Affairs website notes, Fisher decided "to embark on a final career as the Johnny Appleseed of the free-market movement."

Within a few years he'd helped found the Pacific Research Institute in San Francisco, the Manhattan Institute in New York, and the Fraser Institute in Vancouver—and over the decades these seeds sprouted into a global network of more than 150 far-right think tanks. Fisher dedicated special attention in the early days to the Fraser Institute, spending eighteen months starting in 1975 helping it raise funds, hire staff, and establish its ideological presence in Canada. In that role he worked closely with Patrick Boyle, "a Canadian industrialist who had been disturbed by the accelerating left-wards drift of both federal and provincial politics," the biography reads. "The Institute's first publication—*Rent Control: A Popular Paradox*—became a bestseller and rent control was subsequently repealed in almost every part of Canada."

It wasn't until later in the organization's life that the Fraser Institute decided to attack the science of climate change. But the attacks, when they did arrive, came hard and fast. In 1997, the year Kyoto was signed, the institute published a 180-page book called *Global Warming: The Science and the Politics.* "Most politicians, bureaucrats, environmentalists, and members of the media believe as a proven fact that industrial activity—especially the emission of carbon dioxide—is affecting climate by causing an increase in average global temperatures," a description of the book claims. "But the scientific evidence is mixed and scientists continue to debate both the existence and the extent of human-induced global warming." Following the book's publication, the Fraser Institute "became the primary Canadian channel for global warming denial," according to investigative journalist Donald Gutstein.

Around this time, the organization also began attacking the science linking respiratory disease and cancer to smoking. It held two tobacco conferences in 1999: "Junk Science, Junk Policy? Managing Risk and Regulation" and "Should Government Butt Out? The Pros and Cons of Tobacco Regulation." Health experts questioned at the time why the Fraser Institute would risk its credibility by aligning itself with an industry that at that moment was being clobbered in U.S. courts for lying to the public. Leaked documents showed that the Fraser Institute was short on cash and actively seeking funding from cigarette makers. "The area of risk regulation has interested the Fraser Institute for some time. However, lacking adequate funds we have only been able to progress slowly," the institute's Sherry Stein wrote to British American Tobacco in 2000. "We are hopeful that British American Tobacco will elect to support the Institute with an annual contribution of $50,000."

The Fraser Institute had better luck getting checks from oil and gas companies. Exxon contributed $60,000 in 2003 and another $60,000 in 2004—donations given for work on "climate change," according to the company's tax filings. During those years the Fraser Institute published the aforementioned studies "Greenhouse Gas Reductions: Not Warranted, Not Beneficial," and "The Science Isn't Settled," alongside shorter articles like "The Varying Sun and Climate Change" and "Kyoto Crazy." The goal of these articles wasn't necessarily to be coherent or persuasive. "Collectively the texts are not seeking to 'win' any argument," noted Carleton University researcher Aldous Sperl in his dissertation on climate denial in Canada, "but are instead seeking to create doubt and increase skepticism."

That was evident in a report the Fraser Institute released in January 2007, the goal of which was to attack the Intergovernmental Panel on Climate Change, the world's foremost scientific authority on global temperature rise. "The Earth's climate is an extremely complex system and we must not understate the difficulties involved in analyzing it," the Fraser Institute report noted. "Consequently, there will remain an unavoidable element of uncertainty as to the extent that humans are contributing to future climate change, and indeed whether or not such change is a good or bad thing." Actual climate experts slammed the report. Andrew Weaver, then a Canada Research Chair in Climate Modelling and Analysis and a lead author with the Intergovernmental Panel on Climate Change, described the report as "highly ideological."

Yet that didn't stop prominent Canadian media outlets from running extensive coverage of the Fraser Institute report. The *National Post* devoted an entire section on February 6, 2007, called "Inside the Science," to easily debunkable claims made by Fraser Institute skeptics about the IPCC process. The institute saw coverage like this as one of its major reasons for existing. "We are in the business of stimulating demand for improvements to public policy," it explained in a report entitled "How the Fraser Institute Is Changing the

World." "We endeavor to make the results of our research known to the general public through the media."

Soon the Fraser Institute was getting big checks from Koch Industries. It received $150,000 in 2008 from foundations linked to the company, and an additional $175,000 the following year. "I know the grant from the Koch Foundation is for our international work, but I can't tell you which of the projects that it's funding," Michael Walker, former executive director of the Fraser Institute, later told the *Vancouver Observer*. He confirmed that the institute had been receiving money from Koch Industries for "years and years." (As of 2018, public records indicate that Koch foundations have contributed nearly $1.7 million to the Fraser Institute.)

For this reason, Berman's lawsuit included the Fraser Institute on a list of organizations that receive oil and gas funding "to prop up discredited studies and to disseminate misleading information to downplay the severity of global climate change." His complaint gave as an example a coordinated cross-border attack on a 2004 study known as the Arctic Climate Impact Assessment, which warned that the Arctic is heating "at almost twice the rate as that of the rest of the world." The report was trashed by Steve Milloy, then a contributor to Fox News; undermined by a far-right Washington, D.C., think tank called the George C. Marshall Institute; and slammed by the Fraser Institute, which called the Arctic report "an excellent example of the favoured scare technique of the anti-energy activists."

To an average person it might seem like the report was so flawed and controversial that a diverse set of experts felt compelled to weigh in. But in reality, these attacks on climate science, and many other attacks over the years, all shared a common source: they came from groups funded by Exxon and Koch Industries. Those two companies alone contributed nearly $34 million toward denial organizations between 2005 and 2009, according to calculations from Greenpeace. The Fraser Institute, Berman's complaint alleged, was part of an international echo chamber set up to protect the profits of oil and gas companies.

That is not how the think tank presents itself to the public. "Our mission is to improve the quality of life for Canadians, their families, and future generations by studying, measuring, and broadly communicating the effects of government policies, entrepreneurship, and choice on their well-being," the Fraser Institute explains on its website.

"These are people who pretend to be principled," counters Connor Gibson, a former researcher with Greenpeace USA. While groups such as the Fraser Institute might couch their work in notions of "economic freedom," Gibson said, "the outcome is liberty for corporations to plunder the public good, but not liberty for citizens to be free from the plunder. It's a way to justify exploitation of people by a very few private company executives and has nothing to do with the greater good of the country or the world."

That, in his opinion, was the biggest lie of all. It was the lie that Berman was attempting to expose. The explosive revelation that oil and gas companies and their far-right allies had deliberately misled the public hadn't yet been recognized in court. But it didn't need legal certification to profoundly alter how people saw the climate crisis. Thousands of miles away, across the Pacific Ocean, a young Filipina woman was forever changed when she learned this awful truth.

"I remember being angry every day"

THE FLOODWATERS were receding in Tacloban City and the winds were dying down. But Joanna Sustento felt like the world was crashing down on her. She remembers sitting beside a water tank near her wrecked home, "crying my eyes out, screaming my heart out, I was feeling so alone." Around her it was a like a war zone. There were oily puddles where buildings used to be. Twisted metal and broken concrete. Bloated bodies. Typhoon Haiyan had killed over 6,300 people in Tacloban and across the Philippines. Amid the shock of that sudden and violent loss, she couldn't recognize her surroundings. "I was trying to figure out where I was," she said. "Then I realized what I was looking at. I was looking at my beloved Tacloban—destroyed, wrecked, and heartbroken."

Sustento soon learned that her mother, father, oldest brother, sister-in-law, and nephew were among the deceased. Her brother Julius was alive, but badly injured. Sustento and Julius managed to find each other and eventually made their way to their cousin's home, which was still standing after the typhoon. Over the coming days that November of 2013, they tried to get Julius medical treatment for his injured leg. They walked for hours to the nearest hospital but couldn't get him the help he needed. They later made their way to the airport, where Sustento and Julius were evacuated in a military plane to the city of Cebu. "He got confined in a hospital for a few months," she said. Sustento stayed at another cousin's house during that period. Every few weeks she made the trip back to Tacloban, trying to salvage what she could from her family home.

It felt as though she had died and been reborn as someone else. Before the typhoon Sustento lived a happy and conventional life, one relatively privileged by Filipino standards. She worked at an outsourcing company and ran a small online business. She was saving up money, living at home, and dreaming of one day starting her own company. The times when Sustento wasn't building her career she spent with friends and family. "My only concern was about my immediate circle," she said. But in a single morning Haiyan took all of that away. As she processed the trauma of losing her family and seeing the city she grew up in destroyed, she also struggled with another existential crisis. None of her previous goals seemed to mean anything. The challenge now was how to go on. "I didn't know how I would start over," she said.

As foreign aid organizations descended on the Philippines to help rebuild from the typhoon, Sustento began applying for jobs. She was fluent in Tagalog and English and she was an excellent writer. That summer she got hired as an interpreter for a group helping to rehabilitate rural communities across the island of Leyte. The job took her to parts of her home province that she'd never visited. She found it cathartic speaking with other Haiyan survivors and learning their stories. The contours of a new worldview were coming into focus for her. "That was the turning point for me: realizing that I'm just a small part of a bigger community," she said. "I shouldn't only think about myself anymore."

Sustento still found it difficult to talk openly about her experiences during Haiyan. It was easier to communicate through writing. Sustento wrote a minute-by-minute narrative of that terrible morning and posted it online. She got invited to a conference about climate change and mental health. "Telling my story helped me to process," she said. A documentary filmmaker got in touch and interviewed her. Once the film was released, someone from Greenpeace Philippines reached out to Sustento. They wanted to speak with her about climate change.

Everyone Sustento met could agree that typhoon Haiyan was one of the worst disasters to ever strike the Philippines. But why it happened was less obvious. Some people in Tacloban saw Haiyan as God's punishment for their sins. Sustento quickly rejected that explanation. "Children don't have sins," she said. "If it's God's punishment, what kind of god is he?" Groups like Greenpeace made the case that humans were responsible for what happened. Haiyan registered 8.1 on the Dvorak intensity scale—which is only designed to go up to 8.0—with winds of nearly 195 miles per hour. Scientists who have studied Haiyan think that warmer and higher seas caused by global temperature rise made the typhoon more destructive than it would have been otherwise.

Greenpeace explained to Sustento the concept of climate justice, the idea that global temperature rise is not just an abstract scientific phenomenon but a human tragedy that was landing hardest on communities that had done little to cause the emergency. She wasn't versed in this type of language, but after her experiences traveling around Leyte the concept quickly made sense. "I understood it fully," she said. Sustento became obsessed with learning more about climate justice. "I didn't do anything except read about it," she said. Over time she learned that Exxon and other oil companies had known internally about the "catastrophic" risks of climate change long before the public. "Yes, we all do have a contribution, but it's not as big as compared to these big polluters who have known of the climate risk for many, many years," she said.

In some cases, those companies had even seemed to predict disasters like Haiyan. BP in its 1991 educational video on climate change had warned that "from warmer seas more water would evaporate, making storms and the havoc they cause more frequent." That was around the time executives at Shell were learning that in a warmer global climate, "monsoon rainfall ought to increase also as should the frequency of tropical storms."

But instead of using that knowledge to move the global economy away from fossil fuels and save millions of lives, including those of Sustento's family, those companies spent decades spreading lies about the science and sabotaging solutions. It blew Sustento's mind to imagine executives signing off on that strategy. She saw it as so immoral that she "cannot believe there are really people who exist" capable of executing it. As she learned more, bewilderment turned to fury: "I remember being angry every day."

The oil executives who had quietly learned about the deadly consequences of their business model and then buried the research needed to pay.

VII

Well-Oiled Allies

(2016–2019)

"Stacked with friends"

THE FIRST TIME that Rex Tillerson met Donald Trump at Trump Tower in New York, Tillerson entered through the back door. It was December 6, 2016, and Trump was U.S. president-elect, having triumphed over Hillary Clinton with an election night victory so unexpected that Trump himself was reportedly shocked—he later admitted he "sort of" thought he would lose. Now, Trump had to pick his cabinet, and for weeks he'd been trying to decide who should fill the high-ranking position of secretary of state, an official who represents the United States to the world and implements the country's foreign policy. Former New York mayor Rudy Giuliani was begging in public for the job. Mitt Romney, who'd run against Barack Obama for president as the Republican nominee in 2012, was another contender.

But during that Trump Tower meeting on December 6, Tillerson and Trump "hit it off," one transition official told *Politico*. As CEO of Exxon, Tillerson already had profound international influence. The company sold gasoline, diesel, and jet fuel all over the world and produced oil and gas on every continent except Antarctica. Tillerson projected power and masculinity. "He's totally the Trump M.O.," an official said of Tillerson. "Strong guy." Another official explained that Giuliani and Romney by contrast came off a bit weaker. Trump, the official said, liked Tillerson's "outsized, Texan, can-do swagger."

In Canada's oil sands, people were closely watching the nomination contest. If Trump nominated Tillerson for secretary of state, that "would be viewed with a sense of elation" in Alberta, explained Robert Skinner, an executive fellow at the University of Calgary's School of Public Policy and a former oil executive himself. As the majority owner of Imperial Oil, Exxon had always been a major player in the oil sands. But by 2016, its investments there were growing exponentially, rising to 5.1 billion barrels from 1.4 billion barrels in 2007. The oil sands by that point represented 35 percent of the liquid reserves held by the company across all its global operations. Tillerson, the *Globe and Mail* newspaper reported, "knows the Canadian petroleum industry intimately." If he were picked by Trump, "Canadian oil... would potentially have one of its own in the state department."

Tillerson made it clear he would be a strong advocate for the industry. At a Washington dinner in 2015, he shared a table with Canadian ambassador Gary Doer. The two of them chatted about the oil sands, with Tillerson describing projects capable of tapping billions of additional barrels of heavy oil. "He was very proud of those investments," Doer recalled. During that same dinner, Tillerson spoke out in favor of Keystone XL, a project that would transport oil sands crude to heavy-oil refineries in Texas. "The United States and Canada both need this vital pipeline," Tillerson said at the time. "But approval of the pipeline has been taken out of the hands of experienced career officials, and it has become a tool of political manipulation."

Tillerson was referring to a yearslong grassroots campaign against Keystone XL led by Indigenous communities and ranchers in the Midwest, as well as international climate groups such as 350.org, the organization cofounded by author and activist Bill McKibben. Fusing concerns about an oil spill into the region's Ogallala Aquifer with worries about the pipeline accelerating climate-harming developments in the oil sands—James Hansen memorably called Keystone XL "a fuse to the biggest carbon bomb on the planet"—this coalition had by 2015 convinced President Obama to reject the project. Obama made his decision official that November.

Tillerson was adamant the pipeline still go forward. "Keystone XL would improve U.S. competitiveness, it would increase North American energy security and it would strengthen the relationship with one of our most important allies and most valued trading partners," the Exxon CEO said in Washington at the 2015 dinner. But Tillerson also had self-interested reasons to be in favor. The oil sands industry had grown so fast that there weren't enough pipelines to get all the bitumen it produced to market. Building Keystone XL would help ease that transportation bottleneck, and thereby raise the value of Exxon's investments. "The benefit to ExxonMobil could be measured in the tens to hundreds of millions of dollars each year," a report from Greenpeace calculated. "ExxonMobil also has refineries in the U.S. Gulf Coast region that could benefit from the Keystone XL approval."

During the 2016 election campaign, virtually every Republican contender announced they would reverse Obama's cancellation of the pipeline. Trump himself came out in favor that May, saying Keystone XL should go forward provided that the United States receive "a big piece of the profits." Shortly after election day, Trump's spokesperson, Hope Hicks, announced that Kellyanne Conway, a senior transition adviser who had been Trump's campaign manager, would be visiting the oil sands. Conway had accepted an invitation from the Alberta Prosperity Fund, a right-wing advocacy organization with ties to the Fraser Institute. A personal visit from someone in Trump's inner circle was big news in Alberta. The province was home to many "Canadian business leaders keen to show their support for what is widely viewed as a business and jobs friendly administration," wrote the *Globe and Mail*.

The good news for the oil sands just kept on coming. In early December, Trump officially announced Tillerson as his pick for secretary of state. Trump also nominated Rick Perry, a former governor of Texas who was an outspoken supporter of Keystone XL, to the key position of energy secretary. "Donald Trump's prospective cabinet is now stacked with friends of Canadian oil," gushed a story in the *Calgary Herald*. Tillerson apparently wasn't all that concerned about the climate impacts of a pipeline proposed to transport over 800,000 barrels of oil per day through the U.S. heartland. He had earlier called the consequences of global heating causing by the burning of his company's products "manageable," saying humans "have spent our entire existence adapting. So we will adapt to this."

By early January 2017, excitement was mounting in the oil sands for Trump replacing Obama. The Alberta fundraiser that Conway was scheduled to speak at just weeks before Trump's inauguration sold out in days. However, refunds had to be given out when she canceled at the last minute. "I can tell you that Ms. Conway's office and our own have tried every option at our disposal to make this work and, unfortunately, a visit is just not possible at this time," said Alberta Prosperity Fund president Barry McNamar. That hardly dampened the mood in Calgary boardrooms, though. On Trump's first day as U.S. president, he signed an executive order approving Keystone XL. The oil sands pipeline from Canada, Trump later declared, was an "incredible" thing.

"Friends in unexpected places"

IT WAS A MOMENT practically designed to go viral. As a military plane landed at Pearson International Airport in Toronto carrying 163 refugees from Syria, Canadian prime minister Justin Trudeau was there to welcome them. The refugees came off the plane and Trudeau, looking clean-shaven and youthful, kneeled to say hello to a small girl. Photos of this moment, taken shortly after Trump said while campaigning for president that he'd ban Muslim refugees from entering the United States, made international news. "Canadian PM Justin Trudeau shows the world—and Donald Trump—'how to open your heart' as he greets Syrian refugees," declared a headline in the U.K. *Daily Mail*. Later, when Trump made his Muslim ban official, creating chaos and confusion, Trudeau tweeted a photo of himself with the Syrian child with the hashtag #WelcomeToCanada. Within hours, the prime minister's tweet had been shared more than 224,000 times.

These carefully choreographed media events seemed to communicate to the world that Trudeau and Trump had vastly different worldviews. Trudeau was the handsome face of a progressive future, and Trump an ugly reactionary past. But in reality, the two leaders shared more in common than most people were aware. When it came to the crude produced in Canada's oil sands, the Trudeau government actually saw Trump becoming president as a good thing.

"The swearing in of a new administration in the United States that recognizes the strategic importance of Canada's role in North American energy security is, so far, positive news for the Canadian energy sector with regard to a potential increase in energy trade," explain briefing notes for senior officials in Canada's Ministry of Foreign Affairs from May 2017, just months after Trudeau's viral tweets. But the assessment of Trump as "positive news" was not the impression Trudeau's government was giving off publicly. When Trump pulled the United States out of the Paris climate treaty in June that year, a "deeply disappointed" Trudeau released a statement saying that "Canada is unwavering in our commitment to fight climate change."

This was not a one-off occurrence. Saying progressive things publicly while making hard-nosed economic calculations privately was a central feature of Trudeau's political identity. Confusing the public about his true aims is what had allowed him to get elected as prime minister in the first place.

On October 19, 2015, Canadians voted in a leader who claimed to challenge the status quo. "My friends, we beat fear with hope.

We beat cynicism with hard work. We beat negative, divisive politics with a positive vision that brings Canadians together," Trudeau told cheering supporters in Montreal after winning a large victory over Stephen Harper. The forty-three-year-old Liberal leader was the second-youngest person to become prime minister in Canadian history, and his victory ended nearly ten years of Conservative rule. Throughout the campaign Trudeau portrayed himself as Harper's ideological opposite—and on no issue was this more pronounced than climate change. While the Conservative leader called emissions-reducing policies "job-killing," Trudeau declared that "pretending that we have to choose between the economy and the environment is as harmful as it is wrong."

For Canadians exhausted by a government that appointed climate deniers to key positions and prevented its own federal scientists from speaking to the media, the joy of getting Harper out was intoxicating. "Many progressives felt tremendous relief," said Tzeporah Berman, a well-known environmental campaigner who knocked on doors for Trudeau during the lead-up to the election. The night of Trudeau's victory, Greenpeace Canada put out a statement saying that the new government "has an unprecedented opportunity to reject boom and bust polluting industry by stopping tar sands expansion and making Canada a leader in renewable energies." Now was the time, the group's then executive director Joanna Kerr said, for "taking bold action on climate change."

In Canada and around the world it seemed as though Trudeau was poised to deliver. *Time* reported that the Liberal victory was "good news for the fight against climate change," explaining "the defeat of a high-profile climate skeptic like Harper plays an important role in the global debate." The BBC reported that "the environment was one of the top issues for Canadian voters in this election" and that "decisive action" could be on the way. Whereas Harper's unconditional support for the oil sands had given "fossil fuel-dependent emerging economies an excuse to lag on climate control policies," *Foreign Policy* argued, Trudeau's "progressive environmental views," which included a plan to make polluters for the first time ever pay a price for their carbon emissions, "could affect a tectonic shift in global climate politics."

But in reality, as the ensuing months and years would demonstrate, any shift that took place was mostly in people's perceptions. For even as Trudeau talked a big game about addressing the climate emergency, the denial apparatus built up around the oil sands was as powerful as ever. Its priorities both shaped and complemented the federal Liberal worldview. By the time Trump came into office, Trudeau had approved a major oil sands pipeline and greenlit massive natural gas terminals. Bitumen production reached record highs, and within two years, Canada's emissions were thirteen megatons higher than at the peak of Harper's federal majority. Trudeau hadn't closed the gap between the environment and the economy—in some ways he'd made it bigger.

Trudeau became leader of the Liberal Party in September 2013, two years after the centrist party had suffered a huge electoral defeat to the Conservatives. Though Trudeau had portrayed himself as young and idealistic during the Liberal convention that decided the party's leadership, his predecessor Michael Ignatieff saw a "political animal" with a keen understanding of power. "I've got a lot of respect for his political shrewdness," Ignatieff said at the time. "I think he's going to go far."

Less than one month after becoming leader, Trudeau made his first trip to Washington, D.C., where he was hosted by a left-leaning think tank called the Center for American Progress. Trudeau took the opportunity to explain his "steadfast" support for Keystone XL. This apparently surprised some of the people gathered to hear him speak, given that more than a thousand climate activists had been arrested outside of the White House protesting the pipeline over the previous few years, including former NASA scientist James Hansen, who'd claimed the project could be "game over" for the climate.

"There were some people who raised an eyebrow, absolutely," Trudeau said following the Center for American Progress talk. "I'm seen as a strong, young progressive with an environmental background. The fact that I'd be talking positively about the project I think got people thinking about the fact that perhaps it's not as bad as it's been caricatured."

Five days later, Trudeau flew to Calgary and gave a speech to a room of oil executives. "Many of you will know that last week in Washington, I told an audience of American liberals that I support the Keystone XL pipeline," he told the Calgary Petroleum Club. "On balance, it would create jobs and growth, strengthen our ties with the world's most important market, and generate wealth." Trudeau suggested to the crowd that oil and gas companies were having a hard time building projects like Keystone XL because of the aggressive climate change denial of Prime Minister Harper. "After eight years, here is what the so-called friendliest government that the Canadian energy industry has ever had has accomplished," Trudeau said. "The oil sands have become the international poster child for climate change. The government has failed to move the yardsticks on one of the most important infrastructure projects of our generation, the Keystone XL pipeline."

"It isn't working for you," Trudeau said. Under a Liberal government, he explained, Canada would put a price on carbon pollution. This would have the effect of reducing greenhouse emissions. But more importantly, he argued, it would defuse environmental opposition to the oil sands and allow more pipelines to be built. "Let me be clear on this," Trudeau said. "If we had stronger environmental policy in this country: stronger oversight, tougher penalties, and yes, some sort of means to price carbon pollution, then I believe the Keystone XL pipeline would have been approved already. It is the absence of strong policy that makes us an easy target." Trudeau went on: "If we don't convince the world that we have our act together, as a country, on the environment, we will find it harder and harder to get our resources to world markets."

"Keep an open mind," the aspiring prime minister said in conclusion. "You can find friends in the most unexpected places."

You might think speeches like this would have disqualified Trudeau from receiving the support of Canada's environmental movement, but when an election was called two years later, most mainstream groups seemed to shrug it off. Their main goal at the time was getting Harper out of office, an achievement they believed would have global climate implications. Summing up the mood, the media outlet *Grist* wrote that Harper oversaw "an estimated 170 billion barrels of the world's dirtiest oil and [would] do seemingly anything to get it to market."

After the dark years of Harper, having Trudeau campaign for prime minister while saying things like "Canada needs to show the world that it is serious about addressing carbon emissions" was the hope many climate activists desperately craved. Trudeau at one point tweeted a photo of himself in a canoe on the Bow River in Calgary. "He was presented as this kind of river-paddling environmental Adonis," said Martin Lukacs, an investigative journalist and author of *The Trudeau Formula: Seduction and Betrayal in an Age of Discontent*. "He promised that fossil fuel projects wouldn't go ahead without the permission of communities. But the Liberals create these public spectacles of their bold progressiveness while they quietly assure the corporate elite that their interests will be safeguarded."

It wasn't hard to see that in action. One of Trudeau's top policy advisers leading up to the election was Cyrus Reporter, a lawyer with the public relations and lobbying firm Fraser Milner Casgrain who had previously lobbied on behalf of BP and the oil sands producer Nexen. Trudeau's campaign cochair was Dan Gagnier, who even as he worked on the Liberal campaign was providing strategic advice to the pipeline company TransCanada, the potential builder of Keystone XL. TransCanada was also at that time trying to win federal approval for a 1.1-million-barrel-per-day oil sands pipeline from Alberta to Eastern Canada. Days before the 2015 election, Gagnier resigned from the Trudeau campaign after it was revealed he'd "sent a detailed memo" to TransCanada "advising officials there how to deal with a new government," *Maclean's* magazine reported.

Yet climate groups across the country celebrated as the election results came in. "When Mr. Harper's Tories were defeated, there was a period of euphoria and a laying down of arms among activists," the *Globe and Mail* later wrote. A handful of activists kept mobilizing, however. The day that Trudeau moved into the prime minister's residence at 24 Sussex Drive in Ottawa, climate activists organized by the group 350.org and others presented their demands outside the front door. "Number one," Clayton Thomas-Muller explained, "that we freeze the expansion of the Alberta tar sands."

But most such groups scaled back their activity. Lukacs recalls reporting from the 2015 climate change negotiations in Paris—where Trudeau famously proclaimed that "Canada is back"—and witnessing "the leadership of the mainstream environmental movement just getting wasted at wine and cheeses with the Liberal government." Tzeporah Berman saw colleagues taking high-ranking positions in the new government. Marlo Raynolds, former executive director of the clean-energy think tank the Pembina Institute, became the chief of staff for Canada's new environment minister, Catherine McKenna. "I think a lot of us had huge hopes," Berman said. What happened next, she thinks, "is a really important lesson... It's a mistake that access equals influence, which it absolutely does not."

First, Trudeau approved a $1.26 billion liquid natural gas plant on the west coast of British Columbia. "The ardent environmentalists who supported Trudeau, and who passionately oppose the project, are still shaking their heads," journalist Michael Harris wrote at the time. "There's good reason for environmentalists to feel betrayed." Months later, Trudeau signed off on the Trans Mountain pipeline expansion, which would result in nearly 600,000 additional barrels of oil sands being exported from the West Coast of

Canada each day. "The decision we took today is the one that is in the best interests of Canada," the prime minister said. "It is a major win for Canadian workers, for Canadian families and the Canadian economy, now and into the future."

Members of the Tsleil-Waututh First Nation didn't see it that way. Its traditional territories include the area of Burrard Inlet in Vancouver where oil sands tankers would pass by each day—an oil spill, if one someday occurred, would paint the coastline black. "[We are] still of the mind to say no to the expansion," chief Maureen Thomas said at the time. "For me, it means the survival of Tsleil-Waututh Nation for many years to come." Greenpeace, which had cautiously celebrated Trudeau on election night, now argued that the pipeline approval "has broken his climate commitments, broken his commitments to Indigenous rights, and has declared war on B.C."

Berman came to believe that the country's environmental movement had let down its guard at a crucial moment. "You don't build a major piece of infrastructure for a year or two, you build it for thirty or forty years," she said. "While we were all focused on carbon prices and emissions reduction, the fossil fuel industry had been planning to dramatically expand." Lukacs agrees; "Had there been a groundswell of mobilization when Trudeau first came into office, I definitely think the Liberal government would have had to give a whole lot more than they did."

Instead, the oil sands industry continued to rapidly expand. By the end of 2016, U.S. and Canadian producers were planning to nearly double bitumen production in northern Alberta. When you added this to the emissions resulting from gas projects Trudeau was approving, it would make Canada's Paris climate change targets unachievable, according to a report from energy analyst David Hughes. The prime minister who said he was "deeply disappointed" in President Trump for trashing the Paris climate treaty was quietly trashing it himself.

"It just kept going and going"

SCOTT PRUITT was apparently a terrible tenant. The Environmental Protection Agency administrator appointed by president Donald Trump was only supposed to stay in a Washington, D.C., condo owned by a prominent oil and gas lobbyist until he got settled in the capital. It was a sweet deal for Pruitt, who only had to pay $50 per night for a home in an area where places can rent for $5,000 per month. But the former attorney general from Oklahoma, who had won the enthusiastic support of the Koch Industries network for aggressively fighting on behalf of oil and gas companies, just wouldn't leave. "The original arrangement was that he would be there living out of a suitcase... and it just kept going and going," one person with knowledge of the situation told *Politico*. Pruitt was late paying his rent. Eventually, Vicki and Steve Hart, the condo's owners, kicked him out and changed the locks.

But the hassle of having Pruitt crash at their place was potentially worth it in at least one respect. During the time when Trump's EPA head was living there on the cheap, he approved a major oil sands pipeline expansion proposed by Enbridge, the Canadian company based in Calgary. Enbridge at that time was being represented in Washington, D.C., by the lobbying firm Williams & Jensen. And who was the chairman of Williams & Jensen? Steve Hart, who became a reluctant landlord to Pruitt along with his spouse. Enbridge, Pruitt, and the lobbying firm all denied there was any connection between the $50-per-night living arrangement and Pruitt's approval of the Alberta Clipper pipeline expansion, which would bring more than 800,000 barrels of oil sands crude into the United States each day.

Ethics experts were not reassured. "Entering into this arrangement causes a reasonable person to question the integrity of the EPA decision," Don Fox, a general counsel of the Office of Government Ethics during the Obama and George W. Bush administrations, told the *New York Times*. The damage of a pipeline that would stretch from Alberta into the Midwest would be atmospheric as well as ethical. "Alberta Clipper opens the door to explosive development of the nearly 175 billion barrels of tar sands reserves," the Sierra Club claimed in a report on the project. "Analysts have said that to have a chance of limiting global warming to no more than two degrees Celsius, we must leave 66% to 80% of proven fossil fuel reserves in the ground. In stark contrast, over the next fifteen years, the tar sands industry expects to almost triple production of tar sands, the world's most carbon-intensive crude."

As news of Pruitt's condo arrangement with the Washington lobbyists spread, green groups put up satirical posters of Pruitt all over the U.S. capital. "Live luxuriously for cheap—just like Scott," one poster read. Friends of the Earth eventually took credit. "Americans are fed up with Scott Pruitt. He's wasting their tax dollars on his luxurious lifestyle, giving handouts to corporate polluters and poisoning our air and water," said Lukas Ross, the group's lead climate and energy campaigner. With pressure mounting for Pruitt to resign, far-right organizations linked to Koch Industries—including the Heritage Foundation and Americans for Prosperity—launched a campaign to defend him. On Twitter, the Tea Party Patriots urged its over 190,000 followers to retweet a statement calling Pruitt "a leader in rolling back former President Obama's job killing regulations."

It was a noticeable shift in tone from only a few years earlier, when the sprawling political network led by Charles and David Koch had publicly refused to back Trump's campaign for president against Hillary Clinton. "At this point I can't support either candidate," Charles told a gathering of donors. The Kochs weren't pleased with Trump's proposals to ban Muslims from entering the United States and register them in a database, calling them "reminiscent of Nazi Germany." Trump snapped back, ridiculing Charles and David as "globalists," a slur among Trump's supporters. "The good news is that Donald Trump doesn't need the Koch brothers, and he can do this perfectly without their assistance," a Trump campaign manager explained at the time.

That's how it might have appeared to Trump, a billionaire reality TV celebrity who claimed to be self-financing his campaign, but some experts on the Kochs say the political network the brothers spent decades creating and financing was key to Trump eventually becoming president. During the election season itself, the Kochs spent millions of dollars on anti–Hillary Clinton ads in swing states such as Wisconsin and Pennsylvania. Americans for Prosperity, meanwhile, deployed hundreds of staffers across the country to help turn out Republican voters. But more important than that, according to investigative journalist Lee Fang, who was one of the first to link Tea Party protests in the late 2000s to the Koch network, was the "cutthroat" political strategy rooted in big-money disinformation the Kochs had earlier taken mainstream.

Fang saw examples of that strategy in Americans for Prosperity operatives urging Tea Party followers to scream at Democrats during town hall events on health care and capturing the results on video to create viral moments. He also saw it in the Kochs giving hundreds of millions of dollars to think tanks, advertising efforts, and political campaigns disputing the science of climate change, an effort that helped slowed policy action to a crawl, polarized a formerly neutral issue, and protected the Kochs' vast network of oil, gas, and coal operations from regulation. Fang later called David "a fossil fuel industry billionaire" who "spent the second half of his life building a political power structure alongside his brother Charles that radically reshaped society and set the conditions for the rise of President Donald Trump."

Trump soon returned the favor. Despite Trump referring to the Koch brothers while in office as "a total joke," his administration recruited heavily from their network. Trump named Betsy DeVos, a major Koch ally, as secretary of education. Mike Pompeo, another close ally, was appointed new director of the CIA. Trump's choice for vice president, Mike Pence, was reportedly one of the Koch brothers' "favorite politicians," according to a feature by Jane Mayer in the *New Yorker*. But one of the most financially valuable appointments Trump made was installing Pruitt as head of the EPA.

In his previous position as attorney general of Oklahoma, Pruitt aggressively defended oil and gas companies against regulations and investigations. Pruitt signed his name to a letter criticizing federal EPA monitoring of natural gas drilling, not mentioning that the letter was authored by an oil and gas company in Oklahoma called Devon Energy, which also had substantial investments in Canada's oil sands. On another occasion, in 2016, Pruitt coauthored an opinion piece in the *National Review* saying there was nothing wrong with companies like Exxon disputing the science of climate change. "That debate is far from settled," he wrote. "Scientists continue to disagree about the degree and extent of global warming and its connection to mankind." Pruitt made many friends in the Koch network. A private email to Pruitt from an executive at Americans for Prosperity gushed that the attorney general worked "for true champions of freedom and liberty!"

Once installed as administrator of Trump's EPA, Pruitt went to work implementing a policy agenda directly benefiting owners of oil sands projects and other fossil fuel operations. The Alberta Clipper approval was just one piece of it. Pruitt, a longtime supporter of the Keystone XL pipeline, directed the EPA to formally repeal its Waters of the U.S. regulations, a potentially expensive nuisance for pipeline companies wanting to build projects that traverse waterways. His EPA also began the process of formally repealing Obama's Clean Power Plan, a policy pushing the United States toward cleaner forms of energy that was staunchly opposed by Koch-supported organizations. On top of that, Pruitt rolled back fuel-efficiency standards that Koch Industries and other oil companies feared could reduce demand for their products. A spokesperson for Koch Industries suggested to the *New York Times* that the standards were a form of government intervention "that rig the system."

That was consistent with David Koch's own self-serving views on climate regulations. "I think regulating CO_2 excessively is really going to damage the economy," he told Fang in a rare interview from 2010.

Even with public friction between the Kochs and Trump, the billionaires were thrilled by what the president had been able to accomplish in such a short time period. "We have seen progress on many regulatory priorities this Network has championed for years," declared a briefing shared in 2018 during a meeting of a group known as the Koch Seminar Network. "Continuing to remove the burden caused by unnecessary and harmful regulations imposed during the Obama administration—at both the state and federal level—remains a key focus of the Network's overall strategy." The note named as one "notable development" Trump's approval of the Keystone XL pipeline.

By the time the Kochs circulated that note, however, Pruitt was on his way out. The scandal caused by his stay in the luxury condo proved too much even for Trump. Pruitt announced his resignation that April. But in his short stint leading the EPA, Pruitt had proved immensely valuable for Koch Industries, and the oil sands industry that helped bankroll the company, putting in motion regulatory rollbacks and repeals that stood to save the company billions of dollars. It was overlooked during the chaos of Trump' first year, but Pruitt had also given the Koch brothers a gift that couldn't be measured in dollars.

In April 2017, during the period when Pruitt was living out of his suitcase at the lobbyists' home, the EPA gave to Koch Industries its "partner of the year" award. The environmental honor, which was presented in a ceremony in Washington, D.C., was given specifically for energy-efficiency improvements at the company's Pine Bend Refinery, among the largest single refiners of oil sands crude in the

United States. Despite the refinery being one of the biggest emitters of greenhouse gases in all of Minnesota, Pruitt's EPA recognized "Koch's commitment to being a responsible corporate steward," according to a press release. The award Koch Industries received that evening, the company was "proud" to point out, was "the highest honor bestowed by the EPA."

"This is an avalanche"

IT WAS AN OVERCAST September day when lawyers for the cities of San Francisco and Oakland stood at the edge of the San Francisco Bay and announced they were going to sue Big Oil. "These companies knew fossil fuel-driven climate change was real," said Oakland city attorney Barbara J. Parker. "They knew it was caused by their products and they lied to cover up that knowledge to protect their astronomical profits. The harm to our cities has commenced and will only get worse. The law is clear that the defendants are responsible for the consequences of their reckless and disastrous actions." There was a familiar person visible behind her. It was Seattle attorney Steve Berman.

It had been five years since a federal judge threw out Berman's lawsuit on behalf of the Kivalina villagers forced to relocate because of climate change. But now, in the fall of 2017, he was ready to sue Big Oil again. Berman's firm Hagens Berman was helping represent the cities of Oakland and San Francisco in a legal showdown against Exxon, Shell, Chevron, ConocoPhillips, and BP. The cities were facing the prospect of having to spend billions of taxpayer dollars on seawalls and other defenses against rising water. Because that ocean rise was caused by carbon pollution linked to those five companies, and because those companies had a long history of spreading doubt and uncertainty about climate change, Berman thought he could force them to help pay the costs of protecting the Bay Area from global heating.

It was not a great time to be going on the offensive against climate change. Donald Trump had been inaugurated only eight months prior. The president's appointment of former Exxon CEO Rex Tillerson and Koch network stalwart Scott Pruitt to his cabinet signaled an administration that would be friendlier to oil and gas than any other in history. Trump spent much of the first year in office repealing environmental regulations and approving polluting projects. Earlier that summer, Trump withdrew the United States from the Paris climate agreement. Berman saw a lawsuit against Big Oil as one way to fight back. "We have the federal government claiming there's no climate change," Berman explained. "[We can] use the law to accomplish what politicians won't do."

The world had changed a lot since Berman had visited Kivalina in northwestern Alaska and seen waves lapping dangerously high on its shores. Back then, the permanent transformation of the Earth's atmosphere had been widely viewed as a serious but somewhat distant threat. By 2017, it was becoming impossible to ignore. That September, Hurricane Harvey pounded into the Texas Gulf Coast, flooding Houston, causing $125 billion in damage, and killing sixty-eight people. That was followed by Hurricane Maria, which dumped torrential rains on Puerto Rico's mountains, causing deadly mudslides, while 175-mile-per-hour winds raked the island, knocking out electricity for months. More than three thousand people died. Meanwhile, British Columbia had its worst-ever wildfire season. In California, record out-of-control fires burned through tens of thousands of acres. "It was raining fire from the sky," one evacuee said.

City planners in San Francisco and Oakland believed the damage was just getting started. A 2012 report from the California government predicted that the Bay Area's average temperature could rise by more than four degrees Celsius (7°F) over the coming decades. The San Francisco coast had already experienced a sea level rise of almost eight inches. That could triple by mid-century. As waters rose, high-intensity floods that normally would occur only once every hundred years were predicted to occur once per year. By 2100, floods could be inundating the city twice per week, submerging subway tunnels, knocking out utilities, and impacting hundreds of thousands of people. In order to ensure that the city could still remain habitable, officials were planning upgrades to San Francisco's seawall that could exceed $5 billion. Across the city, approximately $49 billion in private and public property was at risk. And that wasn't even including the comparably high numbers for Oakland.

The conventional wisdom was that this was a problem for the Bay Area cities to figure out on their own. Even the heaviest emitters of greenhouse gases are only responsible for a fraction of the global problem. Producers of fossil fuels like oil and gas didn't pollute the atmosphere directly. They sold products to people and businesses living in cities like San Francisco, who then did the polluting themselves from cars, homes, and buildings. "In the case of global climate change, a molecule of carbon is literally around the world in seven days," Scott Segal, an attorney who defends oil and gas companies, told the *Washington Post* in 2017.

But as with Kivalina, Berman believed that the lies told about climate change by Exxon, Shell, Chevron, ConocoPhillips, and BP made those specific companies responsible. If they hadn't spent years misleading the public and lobbying aggressively to block and destroy climate regulations, cities like San Francisco and Oakland wouldn't be facing such a dire emergency. This was the same line of attack that Berman had used in the 1990s against the major cigarette companies. The legal complaint his firm served to the five big oil companies in September 2017 made the link explicit: "Defendants stole a page from the Big Tobacco playbook."

Legally speaking, some observers thought the argument contained in this new lawsuit might not be such a stretch. "The parallels to tobacco are quite clear," said Carroll Muffett, the president of the Washington, D.C.–based Center for International Environmental

Law. Still, nobody had won a lawsuit like this and the oil companies didn't seem to think Berman's latest effort would be any different. "Should this litigation proceed," a spokesperson for Chevron said in reaction, "it will only serve special interests at the expense of broader policy, regulatory, and economic priorities." Less biased observers also had doubts. "I'm not that sure actually," University of Calgary law professor Martin Olszynski told *VICE* in 2017. "It seems obvious that at some point the defendants will say, 'But look, there was demand for our product.'"

Berman had reason to believe San Francisco and Oakland could be more successful. He had tried to fight the Kivalina case in federal court. In retrospect that ended up being a mistake, because judges argued that regulating greenhouse gases federally was a job best left to Congress and the White House. This time around, Berman's legal team was pushing to have the cases heard in California state court, where he predicted the lawsuit would be less likely to raise political questions outside of the court's jurisdiction. And, Berman added, "Some people believe that there are more conservative judges on the federal bench." Meanwhile, the defense team for the oil companies was fighting to get the new suit heard federally. Berman predicted the case was "going to be mired down for the next three or four months in a procedural battle."

There had also been big leaps in climate science since Kivalina. In his complaint, Berman referenced the work of Richard Heede, an independent researcher from California who'd calculated that close to two-thirds of greenhouse gases emitted over the last 150 years could be traced back to just ninety companies. Exxon, Chevron, BP, Shell, and ConocoPhillips were all in the top ten. "We have better science," Berman argued. "We think causation will be easier to prove." And then there was the fact that San Francisco and Oakland were treating climate change as an imminent emergency and beginning to quantify the potential damages in precise financial detail. "The major coastal cities in America and abroad have people worried about this," he said. "That's a big advancement since the Kivalina case."

But arguably the biggest advantage Berman now had was a trove of incriminating oil industry documents. Ever since investigative journalists at the *Los Angeles Times* and *Inside Climate News* had published exposes in 2015 revealing that Exxon had internally researched global temperature rise in the 1970s, new documents continued to surface. A report released in November 2017 by the Center for International Environmental Law contained the revelation that the American Petroleum Institute knew about the dangers of its members' products in 1968, warning that "there seems to be no doubt that the potential damage to our environment could be severe." Another document showed an Exxon researcher stating in 1981 that the company's carbon emissions could "produce effects which will indeed be catastrophic (at least for a substantial fraction of the earth's population)."

These documents were legal game-changers, in Berman's opinion. He referenced them directly in the complaint. "Defendants, both individually and collectively, are substantial contributors to the global warming-induced sea level rise," it read. "Defendants each should have known that this dangerous global warming with its attendant harms on coastal cities like San Francisco would occur before it even did occur, and each Defendant in fact did have such knowledge." If the companies had acted on that knowledge, the complaint argued, shifting to greener forms of energy that didn't destabilize the climate, then we would not be in our current emergency. "Most of the carbon dioxide now in the atmosphere as a result of combustion of Defendants' fossil fuels is likely attributable to their recent production—i.e., to fossil fuels produced by Defendants since 1980."

That was right around the time that the oil sands started to ramp up. If companies like the Exxon subsidiary Imperial had taken seriously the dire warnings from its own scientists, Berman's argument suggested, they never should have started producing bitumen.

Part of the reason for launching the Bay Area lawsuit was to get more incriminating evidence. If a judge decided the case could be heard—a big *if*, considering that no such lawsuit had made it to that stage before—then Berman could call on Exxon, BP, Shell, ConocoPhillips, and Chevron to make internal reports and research into climate change available. "[I'm] dying to get to those documents," he said. "I'm convinced they're going to be smoking hot." More evidence meant more legal vulnerabilities for the oil companies, and more opportunities for attack. "Imagine if I could get ten or fifteen cities to all sue and put the same pressure on the oil companies that we did with tobacco companies and create some kind of massive settlement," he said.

Ultimately that's what this new lawsuit was about. Berman was not just trying to get oil companies to pay for seawalls in the Bay Area. In a broader sense he was attempting to hold them responsible for endangering all human life on Earth. "This is different in kind from anything else," Timothy Crosland, the director of a U.K.-based climate law group called Plan B said at the time. "Once you get started, you get one case that goes through, this is an avalanche. It's got the potential really to bring down the fossil fuel companies."

"We are the beating heart"

ON A SEPTEMBER EVENING in 2019, the premier of Alberta, Jason Kenney, was welcomed like a favorite son at the New York headquarters of the Manhattan Institute. The far-right think tank wasn't a household name by any means. But among conservatives who embraced a hard-line program of slashing regulations on industry while gutting the social safety net, the Manhattan Institute, which described itself as "a leading voice of free-market ideas, shaping political culture," was a very big deal. It had helped shape the thinking of Republican presidents such as George W. Bush ("I thank the Manhattan Institute for all you have done," Bush said in 2008). The think tank also had a long history of spreading disinformation about climate change alongside its Canadian sister organization, the Fraser Institute, at one point referring to human-caused warming as a "myth." Over the years it had received millions of dollars in contributions from Exxon and foundations linked to Charles and David Koch.

That September night Kenney received a gushing introduction. Manhattan Institute president Reihan Salam lionized the Alberta premier's "character" and "largeness of vision and a fearsome work ethic." Thanks to Kenney, the Canadian province home to one of the world's biggest oil reserves was a case study of the institute's ideas put into practice, including, Salam said, "restrained government spending, a more modest tax burden." The conservative movement should be paying attention. "Across the English-speaking world, few conservatives have done as much to adapt time-honored principles to new realities as the premier," he argued.

Kenney could barely contain a smile. "I am really deeply honored by this invitation," he told the gray-haired, well-heeled group. "I grew up as a kid reading Manhattan Institute studies and books and articles." To approving chuckles, he described Alberta as "sort of Canada's Texas. We are the beating heart of free enterprise values in the Canadian political culture. We are the heart of Canada's enormous energy industry."

Amid the smiling faces and warm applause of wealthy and powerful Americans gathered to hear him speak, Kenney radiated confidence. But back home in Alberta, things were not going nearly as well as his New York audience may have assumed. After decades of delay and denial, the disinformation machine built to protect the profits of oil sands producers was starting to smoke and sputter. The industry was facing massive financial threats beyond its control. And the Manhattan Institute–approved policies that Kenney used in an attempt to fix the situation were actually making things worse.

Kenney, who had served in the cabinet of Stephen Harper's "emerging energy superpower" government, had a reputation for being brash and confrontational. But the speech he used to help launch his campaign for Alberta premier in 2018 was especially aggressive. That year's Energy Relaunch conference at the Metropolitan Centre in downtown Calgary was billed as a networking event where oil sands dealmakers and their conservative allies could mingle, talk strategy, and discuss the big question on everyone's minds: "How Canada can regain its competitive edge in international energy markets." Up on stage, Kenney set the tone. Brow furrowed like a Fox News host, he claimed the oil sands industry had become meek in the face of environmental opponents. It was now time to go on the offensive. "I am serious about defending the vital strategic interests of Alberta using every tool at our disposal," Kenney said.

Kenney, the leader of the United Conservative Party, was at that point the front-runner to win a provincial election scheduled for the next spring. He explained that he'd use his position as premier to "fight back" against the province's enemies. Inverting the language people used to critique the Kochs and other billionaires who funded shadowy right-wing causes, Kenney accused climate groups themselves of receiving dark money. The only reason those groups were fighting so hard against the oil sands, Kenney claimed, was because they were being paid by liberal American organizations such as the Rockefeller Foundation. This conspiracy theory had been floating around Canadian politics for years, and it had been thoroughly debunked numerous times.

Yet Kenney vowed to combat foreign-funded "special interests" by setting up a "war room" that would attack anyone spreading environmental "lies" about the oil sands. His government would boycott large U.S. and European banks divesting from the sector for climate and financial reasons. And he would investigate the charitable status of groups pushing for aggressive climate action. While environmental groups were scrambling to comply with intrusive audits, Kenney would instruct oil and gas CEOs to sue them. "If I'm premier we'll be writing checks to allow [oil companies] to go to court," he said. "We'll be supporting pro-development litigation."

Since his childhood days reading Manhattan Institute studies, Kenney had been a keen study of right-wing political trends in the United States. His speech seemed calculated to capture the conspiratorial post-truth zeitgeist remaking American politics under the Trump administration. "This is not energy policy," wrote Alberta energy analyst and journalist Markham Hislop in response to Kenney's speech. "It is Donald Trump-style, grievance-fueled populism pandering to the frustrations and anger of an industry that thinks it is under siege."

On one level Kenney's aggressive plans for Alberta didn't make much sense. Compared to previous years, the oil sands were doing relatively well. The aggregate gross revenue for the five biggest bitumen companies was $140.81 billion in 2018, an increase of 22 percent from the previous year. Shareholders in those companies were also doing great. They received $5.14 billion via dividends and share buybacks, up 83 percent from 2017. The province was getting a decent cut as well: transfers from the oil companies to governments grew by nearly 50 percent during this period. "Oil sands production has never been higher," wrote Ian Hussey, an analyst with the Edmonton-based Parkland Institute.

But by other measures the industry was in serious trouble. The previous year, Shell had announced it would be selling off most of its projects in northern Alberta, a decision influenced by fears that in a world moving to address climate change, the oil sands' high cost and high carbon footprint was a financial liability. (However, Shell still held on to the Scotford refinery it had built to handle oil sands crude in the mid-1980s.) ConocoPhillips sold off $13.3 billion worth of bitumen projects that same year. BP announced it was thinking of leaving. "Companies are finding that further development of the oil sands is just not cost effective," Charles E. Olson, a business professor at the University of Maryland, explained. "More and more, it's also environmentally pretty messy."

The looming threat that neither Kenney nor any bitumen executive could neutralize was a global transition away from oil and gas that seemed to accelerate each year. A week before Kenney's speech to the Relaunch conference, the international consulting firm Wood Mackenzie released a think piece predicting that in less than twenty years, renewables and electric vehicles would become the dominant modes of energy and transportation. "Renewables growth is well ahead of electrification trends but, in time, the two will converge. This is the 'point of singularity,' when the world rings out the old and rings in the new," Wood Mackenzie argued.

These trends were affecting the oil sands. Though they were still profitable, the rapid expansions of the 2000s were long gone. By 2018, there was only one major new bitumen mining project being proposed. Capital expenditure, the money spent by companies to grow the industry, was "not expected to bounce back to boom-time levels and is forecast to further decline in the next decade," Hussey wrote. "This is because . . . the growth phase of the oil sands industry is over." Anticipating this shift, many of the international oil companies were leaving.

But there was a whole world of regional oil and gas companies in Alberta that didn't have that option. Unlike the international companies, which could pick up and go, or the bigger oil sands names like Suncor and Imperial Oil, which were still profitable due to their scale and to decades of trial and error, the mid-size companies, many of which only employed a few dozen people, had slim prospects. They included obscure names like Perpetual Energy, whose share price dropped from $62 in 2014 to less than $0.30 in 2019. The heads of these companies were becoming bitter and conspiratorial. They directed their anger where they believed it could make a difference: against anti-oil "extremists" they alleged were kicking the industry while it was down.

Michael Binnion, head of a struggling gas company called Questerre, urged people to use newspaper reports explaining the mounting climate dangers of fossil fuels "as fire starter to light the coals in your outdoor Hibachi barbeque." An organization he helped found called the Modern Miracle Network attacked the 2015 Paris climate agreement as a "fairy tale" that in reality represented "a massive transfer of wealth under the guise of carbon reductions." President Trump, Binnion's group argued, "wisely announced that America would withdraw."

Paul Colborne, CEO of a beleaguered Calgary oil producer called Surge Energy, summed up the grievances of this cohort and their sense of declining status in an eight-page "statement of facts" he put out as a press release in 2018. "Basically," Colborne said, "after being an extremely ethical, environmentally safe, highly regulated, resource-based economy for the last 100 plus years, why are we suddenly acting 'ashamed' or embarrassed of Canada's abundant crude oil... ?" Surge's share price at that time was just one-eighth of what it had been four years earlier.

When Kenney outlined his "fight-back" strategy, these were the types of people he was speaking to. "The juniors and midcap producers are suffering mightily. Many have failed, a large number of those remaining are highly exposed to price differentials and still losing money, and they are barely hanging on," wrote Hislop, the Alberta journalist. "Kenney represents their grudges, just like Trump represents the grudges of his political base."

Kenney was not fond—at least in public—of these comparisons. "I've been clear about my own political views about Donald Trump," he said. The aspiring premier disagreed with the U.S. president's economic protectionism. The previous year, Kenney had called Trump's Muslim ban an "act of demagogic political theatre." But Kenney had no problem when a United Conservative Party candidate named Devin Dreeshen, who'd worked on Trump's campaign, won a 2018 by-election in Alberta. "I think it's actually helpful to have in our caucus an MLA who can get people on the phone in the

U.S. administration, who knows some of them and has worked with some of them," Kenney said.

On election night in April 2019, Kenney's party won a landslide victory over the centrist New Democratic Party of Rachel Notley.

"We Canadians have been blessed with the world's third-largest oil reserves, and an abundance of natural gas," Kenney boasted during his acceptance speech. "But we have been targeted by a foreign-funded campaign of special interests seeking to landlock Canadian energy." This, Kenney claimed, had "caused the flight of tens of billions of dollars of investment from Alberta, and with it the loss of tens of thousands of jobs." Kenney's tone was menacing: "Your days of pushing around Albertans with impunity just ended... We Albertans are patient and fair minded, but we have had enough of your campaign of defamation and double standards."

That tough talk might have been a bit more intimidating if the $30 million "war room" Kenney set up to fight environmental groups, known officially as the Canadian Energy Centre, didn't quickly descend into farce. Its problems began when a U.S. firm called Progress Software threatened legal action, claiming that the Centre's logo was virtually identical to its own corporate logo. Kenney's Centre responded by designing a new logo, a stylized red-lined "a" with a maple leaf on its corner. Within a week another U.S. company was threatening to sue. "This is our intellectual property—that logo is similar to ours," said a spokesperson for ATK Technologies.

Several months later, the Centre again found itself attracting unflattering attention. This time it began with a news feature in the *New York Times* explaining why a growing number of financial institutions—including HSBC, BlackRock, and The Hartford—were choosing to sell off their investments in the oil sands. Kenney's "war room" responded with a series of tweets attacking the *Times* for being "routinely accused of bias" and not being "the most dependable source." With no factual basis for these accusations, the Canadian Energy Centre was forced to backtrack. "I apologize for some of the tweets," the head of the war room, Tom Olsen, said. "The tone did not meet [our] standard for public discourse."

Kenney had promised to make Alberta's economy great again, but his "free-market" strategy was also a flop. Among his first major policies was to cut corporate tax rates by 1 percent, which he offset by effectively eliminating over 1,400 jobs in health care, education, and other areas of the public sector. Kenney insisted this would "renew Alberta as a magnet of job creation and investment." Instead, it boosted the margins for already profitable oil and gas companies. Net incomes of bitumen producers like Suncor and Cenovus grew by an estimated $2 billion. Yet ordinary Albertans saw little in return. After receiving a $55 million tax cut, the oil and gas company Encana relocated to Texas. Husky laid off hundreds of Albertans while saving $233 million in taxes. Overall employment in the resource sector declined throughout 2019 by ten thousand jobs.

"Many of the larger companies with spare cash on hand are using the funds to pay down debt and buy back shares, instead of spending on new projects," the CBC reported.

Kenney did not share this information during the September evening event at the Manhattan Institute. Kenney's hosts, even if they were aware of the growing financial and political turmoil facing the oil sands, didn't ask. Instead, the organization's president, Salam, presented to the conservative New Yorkers in attendance a vision of the oil sands that conveniently ignored the truth. In this alternate reality, a government that spreads baseless conspiracy theories, brings in tax cuts that primarily benefit the already well-off, and refuses to acknowledge a global transition away from fossil fuels was doing a great service to its citizens. "Premier Kenney's many successes in government have benefited first and foremost his fellow Albertans and Canadians," Salam said. "It was they who gained."

The room erupted in applause for Kenney.

"They surrounded me"

ON A DRIZZLY SEPTEMBER DAY in 2019, Joanna Sustento showed up to Shell's headquarters in Bonifacio Global City in greater Manila. Around the world, climate groups were organizing a coordinated climate strike that was predicted to bring millions of young people into the streets of New York, London, Montreal, Mexico City, Nairobi, and countless other cities. But with the tropical rain coming down, Sustento stood all alone in a square outside Shell's building. She had only a loose idea of how things might go. She was holding a sign that said "climate justice" and she had a written message to Shell's CEO: "My community is demanding justice for the thousands of lives killed, for the hopes and dreams of a better future lost to greed, apathy and deceit that the fossil fuel industry supported." Before Sustento could deliver that message, dozens of police officers showed up to arrest her. "It was ridiculous," she said. "They surrounded me and took me to the precinct."

In Sustento's earlier life, before the disaster that killed her family, she never could have imagined being hauled away by police. Her dad had been in the military, and he often made it known that he saw activists as troublemakers who couldn't hold down a real job. "He didn't have the nicest things to say about them," she said. Sustento internalized those views. "You get influenced by your family," she said. "I viewed protesters as people who don't do anything but complain."

But in the years following Haiyan, her perspective shifted. One of the first things she remembers changing her mind was a thousand-kilometer "climate walk" (around 620 miles) organized by Greenpeace and local environmental groups in 2014. Starting in Manila, dozens of activists walked for forty days with the goal of waking up communities "to the impacts of disasters and climate change." When the marchers arrived in Tacloban on the first anniversary of the typhoon, Sustento didn't reflexively judge them as complainers, as she might have done before. Instead, she said, "I was amazed—who were these people?"

The following year, a former climate negotiator for the Philippines named Yeb Saño helped lead a march from Rome, Italy, to the 2015 climate talks in Paris. When the marchers arrived in the French capital in December, Yeb's brother A. G. Saño painted a mural that he called "the face of climate change." A. G. had survived Typhoon Haiyan, but his good friend hadn't. When Sustento saw a photo of the mural, she was surprised to recognize the person it depicted. "The face was the face of my eldest brother, who died during the storm," she said.

Later, Greenpeace asked Sustento if she would share her story of Haiyan on the group's website, and she quickly agreed. Soon she was helping lead direct actions herself. In 2017, Sustento accompanied the actress Lucy Lawless to the Barents Sea, where they were part of a small team that traveled in inflatable boats to an oil rig for the Norwegian state-owned company Statoil. "It is hard for me to grasp and accept that a government like Norway's is opening up new Arctic oil drilling, knowing full well it will put families and homes in other parts of the world at risk," Sustento said. After losing everything, she saw activism as a way to start over. "I felt like I had a second life and I wanted to make something beautiful from what was left," she said.

While she was getting more involved in the cause, Greenpeace East Asia was pursuing an unprecedented legal challenge to the oil and gas companies it deemed responsible for Haiyan. Greenpeace and other NGOs presented their argument in a 2015 petition to the Philippines Commission on Human Rights. They pointed to the work of the California researcher Richard Heede, who had determined that just forty-seven investor-owned "carbon majors" were responsible for nearly one-quarter of all industrial greenhouse gases released into the atmosphere between 1751 and 2013.

Exxon, through its vast global operations, including its subsidiary Imperial Oil's involvement in the oil sands, released 3.52 percent of those gases. Add Chevron, BP, Shell, ConocoPhillips, and Suncor, all with significant bitumen investments, and the number rises to 12.7 percent of a figure that comprises all the industrial emissions ever released since humans started burning fossil fuels. "They have manufactured, produced, and sold fossil fuel products for decades with the knowledge that those products would warm the climate," Heede said. "They have a substantial responsibility for the damages that their products have caused."

The petition submitted by Greenpeace described damages in the Philippines ranging from crop failures to rising seas, and from ocean acidification all the way to disasters like Haiyan. "Climate change interferes with the enjoyment of our fundamental rights as human beings," it read. "Hence, we demand accountability of those contributing to [it]." The Commission on Human Rights reviewed the petition and agreed the evidence was solid. It announced that it would open an investigation into whether the "carbon majors" named in Heede's research had through their polluting business models and disinformation campaigns caused human rights violations. "This kind of case has been filed before other national human rights institutions, but they all rejected it," said human rights commissioner Roberto Cadiz. "But [we] said, 'Alright, we are willing to be the first.'"

That couldn't have been welcome news to the companies named in the investigation. Yet they did their best not to show it. Many of the carbon majors simply ignored a request to provide input. Some listed their investments in renewable energy as proof they were taking the climate emergency seriously. Others effectively tried to shut the investigation down. "We have made various legal challenges to the Commission's activity," said ConocoPhillips, whose oil sands operations include a 148,000-barrel-per-day operation in northern Alberta.

In theory the companies didn't have much to worry about. They were being investigated by an underfunded commission in a low-income country with no binding legal power. But the investigation threatened to destabilize the narrative barricades these companies spent decades building: that the science linking their business models to global temperature rise is uncertain, that each and every one of us is responsible for a global problem, and that anyone who argues otherwise is on a mission to destroy the economy.

That was why the threat of legal accountability—not just in the Philippines but in courtrooms across the world—was beginning to cause internal alarm. "In public, the oil majors dismiss what they see as political theatre, but there is growing concern that scientific work linking hydrocarbons to climate change and climate change to extreme weather could produce judgments against the industry as a whole and perhaps individual groups," argued Nick Butler, a *Financial Times* columnist who spent nearly three decades working at BP.

Sustento was asked to speak at the opening of the investigation. On a muggy afternoon in December 2017, she got up on an outdoor stage in Quezon City and began telling the story of how Typhoon Haiyan killed five of her close family members. In the audience that day were dozens of other people from Tacloban City and the surrounding area. As Sustento addressed the crowd in Tagalog, men and women wiped their eyes, and there was a heavy silence when she paused to check her notes. "Climate change," Sustento concluded in English, "is one of the biggest injustices in human history."

The event was taking place outside of the headquarters of the Philippines Commission on Human Rights. Inside the building, commission representatives, climate activists, and a few local journalists were waiting to see how the day would unfold. The commission had invited every fossil fuel company named in the Greenpeace petition to send representatives. But as the conference officially got started, nobody seemed sure which, if any, of these carbon majors would actually attend.

By the end of the afternoon it was clear that none of them—not Suncor, Exxon, Shell, ConocoPhillips, BP, Encana, Husky Energy, Nexen, or any of the several dozen other oil, gas, and coal companies earning a fortune by destabilizing the Earth's atmosphere—were going to show up. Commissioner Cadiz shrugged it off. "We do not have territorial jurisdiction over these companies," he said, "we can't award damages against them, but that's a very narrow view." Cadiz was already planning additional hearings in the United States and Europe. He was seeking a global audience for a counternarrative on climate change: the emergency was being accelerated by a small group of fossil fuel producers that deliberately lied to the public about the impact of their products.

Sustento was in a good mood by the end of the day, despite none of the fossil fuel companies showing up. "At first I thought it's kind of bad news because it means that they don't care about the people here, or maybe they think it's just a joke," she said. But spending the day with Haiyan survivors and telling her own story had been invigorating. She was convinced the oil and gas giants named in the investigation could only evade justice for so long. "They're underestimating what we can do," she said.

The following year, in September 2018, the Commission on Human Rights held a hearing in New York. It was the same city where, nearly sixty years earlier, Edward Teller had first warned oil executives about the looming dangers of climate change. By now, it was clear that the terrible impacts he'd predicted had arrived. At the 2018 hearing, former UN special rapporteur on human rights and the environment John Knox said that "the Philippines is already experiencing this future through Haiyan and other events." During a hearing several months later in London, England, Henry Shue of the University of Oxford's Centre for International Studies said that it's "quite clear that the harms done by climate change violate human rights."

The commission presented its final ruling at the 2019 climate talks in Madrid. It found that Exxon, Chevron, Shell, Suncor, and dozens of other oil and gas companies had "a clear moral responsibility" for causing climate change. While the ruling itself was a recommendation with few direct legal consequences, Cadiz said it showed that courts in Canada, the United States, Europe, and other places would have no shortage of evidence with which to bring legal charges against the fossil fuel industry. And in cases where companies "have been clearly proved to have engaged in acts of obstruction and willful obfuscation" on climate change, including spreading denial about the science, he saw no reason why they shouldn't face criminal charges.

As environmental scholars and analysts debated legal questions raised by the Philippines investigation, Sustento was attempting to address personal ones. This was around the time she got detained by police in Manila for her protest outside of Shell's offices. The officers took Sustento to a police station for questioning. "They asked for the purpose of my activity," she said. "I told them I just wanted to speak to Shell management." After she was released several hours later, Sustento went back to the Shell building. She had something that she wanted to deliver to the company along with her written message. It was a poster showing the smiling face of a child, one young life among thousands stolen by Typhoon Haiyan. The three-year-old boy on the poster was Sustento's nephew, Tarin.

VIII

The Right to Live

(2020–2022)

"Robbed of their options"

NOT LONG AFTER Lucy Molina moved with her children to the Adams Heights neighborhood of Commerce City, Colorado, she started to develop vertigo and get bad migraines. Soon her kids were getting sick too. Sometimes they experienced nosebleeds and headaches so bad it was difficult for them to go to school. Molina had chosen this neighborhood in Greater Denver for her family because it was relatively cheap by Colorado standards and had a large community of Latino people living there. But as time went by without improvements, she began preparing her children for the worst. Molina taught them how to drive so that if her dizziness ever became too intense to get behind the wheel, they could take her to the emergency room.

It wasn't just Molina and her kids who were ill. In the nearby Latino neighborhoods of Globeville and Elyria-Swansea, people were experiencing asthma at higher levels than in other parts of Denver, according to a Denver Department of Public Health & Environment report. Children and young people in these neighborhoods also visited emergency rooms for asthma more often. When the coronavirus pandemic arrived, these communities were some of the hardest hit in Colorado due in part to their residents' preexisting lung conditions. A record wildfire season that summer made things even worse. Community members couldn't hang out indoors for fear of catching COVID-19. But if they gathered outside, they'd have to breathe wildfire smoke.

To Molina it sometimes felt as though everyone she knew was ill. "When other communities are planning their summer, their vacation to Hawaii or Mexico, our kids are planning how to take care of their parents in case they get sick," she said. But by the summer of 2020, it was no longer a mystery why.

A doctor who Molina saw for her migraines suggested that a large oil refinery operated by the Canadian oil sands company Suncor could be to blame. The facility, which produced 98,000 barrels of gasoline and diesel per day, was located only a few miles away from where she and her neighbors lived. The doctor told Molina to read a Physicians for Social Responsibility report that said exposure to oil facilities can have health impacts including "cancers, asthma, respiratory distress, rashes, heart problems, and mental health problems." A study in the *Journal of the National Cancer Institute* found that Americans living within five to ten miles of an oil refinery were at higher risk of getting several different types of cancer.

Molina became convinced that the polluted air she and her family were breathing—which contained high levels of sulfur dioxide, hydrogen sulfide, hydrogen cyanide, nitrogen oxides, carbon monoxide, and particulate matter from the oil refinery—was what was making them sick.

Suncor didn't address those fears directly in a VICE *News* story about its Colorado facility, instead explaining that "we've invested over $1.3 billion in the refinery in upgrades, technologies, and systems that improve the reliability of our operations." Yet the company acknowledged the accuracy of reporting in the *Denver Post* showing that its refinery had exceeded safe pollution limits for hydrogen sulfide, carbon monoxide, and sulfur dioxide more than a dozen times between March 27 and April 22, 2021.

Such fumes may have increased the risk of people getting severely ill from COVID-19. A Harvard study from April 2020 found that "people with COVID-19 who live in U.S. regions with high levels of air pollution are more likely to die from the disease than people who live in less polluted areas." The wildfires that were like an extra vise squeezing people's lungs were also linked to the refinery. That was because the Suncor facility was one of Colorado's largest carbon emitters, releasing nearly one million tons of greenhouse gases each year. Those gases were in turn helping alter the climate, and thereby increasing the risk of forests going up in flames.

Molina became one of the refinery's most outspoken critics. She joined community members and environmental groups in trying to shut Suncor's facility down. "We're in the heart of it, right by all the polluters, so for me it's an ongoing fight for the right to live," Molina told the local news outlet *Westword*. "If we don't take action, we'll have no world left to fight for."

As Molina and her neighbors mobilized in the streets, a separate legal effort against Suncor was gaining momentum in the courts. In June 2020, the U.S. Court of Appeals for the Tenth Circuit ruled that an unprecedented lawsuit filed in Colorado against Suncor could proceed in state court. The lawsuit was brought on behalf of the city of Boulder and the counties of Boulder and San Miguel. It took direct aim at the Commerce City Refinery, accusing Suncor of emitting "substantial amounts of [greenhouse gas emissions] in Colorado from its fossil fuel operations, including refining and transportation activities," which were increasing the risks of drought, extreme heat, pine-beetle infestations, and other factors contributing to wildfires in the state.

Because 20 percent of the crude oil processed at the refinery was derived from the Canadian oil sands, the complaint read, Suncor's heavy-oil operations in northern Alberta, which "produce a proportionally greater amount of [greenhouse gas] emissions than most fossil fuel companies do," were a major contributing factor to the climate damage Coloradans were experiencing. The lawsuit also named Exxon as a defendant, arguing that it and Suncor were jointly implicated through their partnership in Syncrude, "a large, if not the largest, tar sands developer in Canada, which promotes and sells in Colorado synthetic crude derived from the Canadian tar sands." But the lawsuit didn't stop there. It accused Suncor and Exxon of lying about the climate emergency for nearly sixty years. If those companies had taken seriously the internal scientific warnings about the dangers of their industry starting in the 1960s, the complaint argued, they shouldn't have developed the oil sands at all.

It was no coincidence that this legal effort echoed the lawsuits that Steve Berman had filed on behalf of San Francisco and Oakland several years earlier. The Colorado lawsuit, and more than a dozen others across the United States, were inspired by the Bay Area litigation. Several months after those 2017 cases in California were announced, Santa Cruz and Santa Cruz County sued twenty-nine fossil fuel companies for climate-related damage. Early the next year, New York City filed its own lawsuit, with legal representation from Berman's firm. The city of Richmond, California, sued three weeks after. King County, Washington, which is home to Seattle, demanded climate compensation from BP, Chevron, ConocoPhilips, Exxon, and Shell in May 2018. That was also a Berman case. The lawsuit in Colorado against Suncor and Exxon was filed that summer.

Oil companies dismissed these lawsuits as a form of political theater. A spokesperson for Exxon at one point claimed that legal action brought by Colorado municipalities is "wast[ing] millions of dollars of taxpayer money and do[ing] nothing to advance meaningful actions that reduce the risks of climate change." But some legal experts thought Big Oil was more concerned than it was letting on. "It would be absolutely devastating to the industry to have one of these high-profile cases go to trial," said Mary Wood, a professor of environmental law at the University of Oregon. "Not only for the legal precedent it could create, but just as important, a trial would result in a broad moral indictment of this industry."

But these lawsuits were a long distance from that stage. The judge hearing the Bay Area litigation in 2018 had first wanted a lesson on climate change itself, demanding a five-hour tutorial on the history and impacts of global temperature rise. It featured climate scientists, environmental experts, and representatives for Shell,

BP, ConocoPhillips, Chevron, and Exxon. Afterward, Judge William Alsup seemed sympathetic to the oil companies. "I look at the broad sweep of history and see that we needed oil and fossil fuels, coal would be another one, to get from the 1850s or 1859, when they struck oil in Pennsylvania, to the present," he said. "And yes, that's causing global warming, that's a negative, but against that negative we need to weigh in the large benefits that have flowed from the use of fossil fuels." Not long after, Alsup dismissed the case. But that wasn't the end of it. San Francisco and Oakland appealed to the U.S. Court of Appeals, and in August 2020, that court reversed the ruling, bringing the case back to life.

That was around the time that a separate court dismissed a legal challenge from Exxon and Suncor and ruled that the Colorado case could also move forward. At this point more than twenty cities, states, and other jurisdictions in the United States were taking on Big Oil in court. But the Boulder legal effort was the only one to single out the Canadian oil sands directly. The timeline of corporate denial on climate change that it sketched out began in the late 1950s. That was when Sun Oil executive Robert Dunlop attended the Columbia University symposium where Edward Teller warned of a "greenhouse effect" caused by burning fossil fuels that "is more serious than most people tend to believe."

That wasn't the only warning that predecessors to Suncor received, however. The complaint referenced the 1968 American Petroleum Institute report, detailed earlier in this book, which explained that if society keeps burning climate-warming fossil fuels "the potential damage to our environment could be severe."

As the complaint alleged, there is evidence to suggest Sun Oil received this warning. The president of Sun Company of Canada during this period, Wilburn T. Askew, often served on the American Petroleum Institute's technical committees. Dunlop, the Sun Oil president who listened to Teller speak in New York, was the president of the American Petroleum Institute from 1965 to 1967. "From our perspective, it's not speculative to say that Suncor, through its and Sun Oil's participation in [the American Petroleum Institute] was very much aware of the climate science that was circulating at the time," said Marco Simons, lead counsel with the organization EarthRights International, which was involved in the lawsuit.

But, as the reader will be well aware by now, the company did not appear to take that knowledge seriously. In 1967, Sun Oil and its Canadian subsidiaries officially opened Great Canadian Oil Sands, the first commercial bitumen project in Alberta. This eventually allowed Canada to develop one of the biggest oil reserves in the world. "At the latest, Suncor was told a year later about the dangers of unchecked fossil fuel use, but plunged forward to this day regardless," the lawsuit alleged.

Not only did Suncor ignore a growing body of climate science over the following decades, according to the lawsuit, the oil sands giant was involved in efforts to "confuse the public and consumers about the risks of alterations to the climate from fossil fuel use, in order to maintain fossil fuel demand and their fossil fuel business." The lawsuit claimed Suncor was implicated through its membership in the American Petroleum Institute, which during the 1980s and 1990s ran and directed the Global Climate Coalition that spent millions of dollars on advertising claiming that fears over global warming were overblown and inaccurate. Exxon, meanwhile, was one of the coalition's corporate leaders.

There is little doubt that these disinformation campaigns shifted public opinion. "In 1992, 88 per cent of Americans believed that global warming was a serious problem, but by 1997 that number had fallen to 42 per cent (with only 28 per cent of Americans thinking immediate action was needed)," the lawsuit read.

Attorneys for Exxon argued that the entire premise of the lawsuit was bogus. "Anyone who used automobiles, jets, ships, trains, power plants, heating systems, and factories since the dawn of the Industrial Revolution," the oil company's legal team wrote, "has contributed to the problem of global warming."

Despite the legal uncertainties, the Colorado plaintiffs argued that what they and other climate litigants across the United States were trying to prove for the first time in court was simple: oil sands companies knew about the dangers of climate change decades before the general public, made huge profits selling products that accelerated the emergency, spread doubt and confusion about the science, and then passed off the heavy costs of disasters and extreme weather onto cities like Boulder.

"It is far more difficult to change it now than it would have been if the companies had been honest about what they knew 30 or 50 years ago," Simons, the EarthRights attorney, told the *Guardian*. "That is probably the biggest tragedy here. Communities in this country and around the world were essentially robbed of their options."

Lucy Molina was seeing the impacts of that firsthand. After years of breathing in air pollution from the oil sands, members of her community were being ravaged by the pandemic. "We are more at risk of dying or getting COVID," she said. "My boss is on a ventilator fighting for her life. You become numb after a while but it still hurts." Molina felt like the community's only option now was to fight back. "[We] need to hold Suncor accountable," she said.

As the climate lawsuit against Suncor and Exxon navigated a maze of legal challenges and appeals, other municipalities and states across the U.S. added their names to the legal movement to sue Big Oil. The attorney general of Connecticut filed a lawsuit against Exxon in September 2020. It was followed by the state of Delaware; Hoboken, New Jersey; Charleston, South Carolina; Honolulu, Hawaii; Annapolis, Maryland; New York City; the state of Vermont; and several others. "Communities are tired of paying the price for the fossil fuel industry's climate deception, and it's only right that they turn to the courts for justice," said Richard Wiles, executive director of the Center for Climate Integrity, a group supporting legal action against the oil industry.

In this growing wave Wiles saw a warning for the industry: "Big oil and gas companies can no longer lie to the American public without consequences," he said. He believed that this time around, it was a warning that oil executives would have a much harder time ignoring.

"Why wouldn't I choose the right thing to do?"

AS THE DIRE EFFECTS of the growing climate catastrophe became more evident, and in-depth journalism, scientific testimony, lawsuits, and advocacy campaigns chipped away at the industry cover-up, thoughtful employees of major oil firms began questioning the moral implications of their careers. But it's likely that few experienced a crisis of conscience more profound than Enrique Rosero, a bright, young engineer at Exxon. Alarmed and disturbed by what he was reading about his employer and industry, Rosero decided to test the limits of dissent at a town hall meeting for employees in early 2020. Exxon's swift and harsh reaction, he says, was intended to silence him and all others concerned about the climate emergency.

Rosero was not one to doubt the motives of his employer. Ever since he'd joined Exxon in 2009, he'd found the words of the company's executives comforting. The narrative those executives presented to the public was that a great global energy transition was underway—toward cleaner technologies, yes, but also toward a future in which billions of people in the developing world would need affordable sources of energy to lift themselves out of poverty. Exxon's official slogan at the time was "Taking on the world's toughest energy challenges." "So we were going to do the hardest, most technical things, going to the places where no one has been and do our best to apply knowledge and technology," Rosero said of how he understood the company's mission.

Rosero knew that as one of the world's major oil producers, his employer was contributing heavily to global temperature rise. Yet he also felt that Exxon was in a good position to help address the challenge. A decade earlier, Exxon's then CEO Lee Raymond had aggressively denied the scientific consensus the company's own researchers had helped create, infamously telling attendees at a major petroleum event in China that "the earth is cooler today than it was twenty years ago." Rosero figured the sharp minds inside the company had made the necessary correctives since then. Speaking in 2007, Raymond's successor as CEO, Rex Tillerson, had proclaimed, "Our industry has a responsibility to contribute to policy discussions on these important issues—and to take concrete actions ourselves to reduce emissions."

Rosero had begun his career at Exxon almost by chance, not long after Tillerson began waxing concerned about the climate. At the time Rosero was an international student from Ecuador, doing a PhD in geosciences at the University of Texas at Austin. Oil and gas companies were recruiting at his school: "Several companies came every single day for a week," he recalled. The industry's workforce was growing older, and top producers worried that a wave of baby boomers were about to retire before those workers could pass on their decades of knowledge and expertise to an incoming generation. Rosero walked by the recruiters each day on his way to class but didn't seriously consider what they were offering because his scientific interests lay elsewhere. "I wasn't planning on interviewing," he said.

But Rosero slowly grew curious. One day he decided to check out an information session hosted by Exxon. "I entered the room, it was mostly empty, there was more pizza than people," he said. "I just sat down and listened to what they had to say." Rosero, initially noncommittal, warmed to the recruiters' pitch. "It was kind of interesting because they were talking about technology, supercomputers, new solutions to provide the world with energy. So the next day I took the interview and they invited me for an internship."

Rosero was always interested in science and engineering. He had attended graduate school in Germany and Utah, where he studied environmental fluid mechanics and hydrology and water resources. An accomplished researcher, he got a full scholarship for his PhD. Rosero doesn't recall having strong feelings about climate change at the time. But he does remember a furious reaction at the school in 2008 when some older faculty members invited Fred Singer, the well-known climate change denier whose research had been funded by Exxon and other oil producers, to speak. Students and some professors objected, and in protest they published a statement explaining why climate change was real. "I learned that instead of opposing the views of others," Rosero said, "proclaiming what you stand for and what you believe is the way to go."

What Rosero's new employer really stood for and believed, however, was not as it seemed. Exxon itself may have been done contesting climate science in public, but it was still privately funding deniers to do that work instead. In 2008, it gave $76,106 for a project by astrophysicist Willie Soon entitled "Understanding Solar Variability and Climate Change: Signals From Temperature Records of the United States." Soon, who throughout his career had also received funding from the American Petroleum Institute and the Charles G. Koch Charitable Foundation, had long argued that climate change was less severe than most scientists were predicting. In one paper republished by the Fraser Institute, he claimed that there were many "uncertainties in understanding climatic change, including natural factors."

In 2009 alone, Exxon contributed $1.3 million to at least two dozen organizations attacking the science of climate change, according to Greenpeace calculations. That included $100,000 to a group called the Atlas Economic Research Foundation, which helped organize a conference that year entitled "Global Warming: Was It Ever Really a Crisis?"

Rosero didn't know about any of this when he began his internship at an Exxon research lab in Houston over the summer. He found the work challenging and exciting. "I imagined I could like working for them," Rosero said. He had a year left in his PhD, and when he graduated, he joined Exxon right away. Starting out as a senior research scientist in 2009, he was soon promoted to research specialist. That position involved him providing technology support to geoscientists working at Exxon's subsidiary Imperial Oil. Bit by bit, he was learning about the true ecological impact of his employer.

In the summer of 2012, shortly after his daughter was born, Rosero traveled to Imperial's Kearl oil sands mine in northern Alberta. At twenty thousand hectares, it was larger than the New York City borough of Brooklyn. Flying north from Calgary, Rosero saw vast open prairies give way to the boreal forest, stretching out as far as he could see. "And then you see this black hole fuming with steam," he recalled of approaching the Kearl mine. "It looks like Mordor from the air," Rosero said, referencing the dark realm in *The Lord of the Rings* ruled by Sauron. "That was very shocking for me, like, 'What the heck are we doing?'"

As startling as it was to witness his employer's environmental footprint firsthand, Rosero remained committed to Exxon. As the years passed, he worked his way up to the position of "Upstream Uncertainty Analyst." His job was to quantify the size of new oil reserves found by Exxon and combine that with a profitable plan to produce the oil. He worked with geoscientists, development planners, and commercial analysts. "It was very fun, to tell you the truth; they put top people in the top projects," he said.

Rosero's work helped make possible a development plan for massive offshore oil reserves in the South American country of Guyana, which were estimated to contain over eight billion barrels of oil. Darren Woods, who became Exxon CEO in 2017, called the Guyana operation, one of the biggest oil projects proposed by any oil producer anywhere in the world in the decade leading up to 2020, "an integral part of our long-term growth plans." Horrified onlookers such as the European environmental group Urgewald called the operation "a carbon bomb" that could result in over one billion tons of planet-warming carbon dioxide being released into the atmosphere.

To do this work while also caring about climate change required Rosero to see Exxon as a good-faith developer of global energy solutions. This was emphasized in an updated Exxon slogan known as the "dual challenge," which the company defined as supplying "the affordable and reliable energy the world needs while reducing environmental impacts." Rosero, like many of the employees he knew, internalized this message. He told himself that even under a rapid shift to greener forms of energy, the world would still need oil. Exxon should be the company to provide that oil, he believed, because it possessed the expertise, technology, and global reach to scale up renewables and offset its emissions.

Around the time he was working on projects in South America, Rosero decided to try actually putting that idea into action. He volunteered for a team focusing on the environmental side of Exxon's "dual challenge." Their task was to scale up technology for capturing and storing greenhouse gas emissions produced by the company. Rosero says he helped figure out a plan involving sequestering huge amounts of carbon dioxide underneath the deepwater floor of the Gulf of Mexico. But the team couldn't get money or resources to begin executing the plan. "We didn't have any mandate," he claimed. "It got to a point where there was not going to be a path forward." Rosero politely resigned from the team. (A later investigation by *Inside Climate News* suggested that as of 2020 Exxon was mostly using carbon-capture technology to pump the last dregs of oil out of near-depleted reserves—not address climate change.)

For much of the time he worked at Exxon, Rosero didn't take the company's environmental critics very seriously. It sounded hyperbolic to him that Exxon had led a vast conspiracy to mislead the public about climate change. Unlike those environmentalists, Rosero actually worked inside the company. Many of his colleagues were decent people who cared about the future of the planet. But then came reporting in 2015 from the *Los Angeles Times* and *Inside Climate News* that for the first time described "how Exxon conducted cutting-edge climate research decades ago and then pivoted to work at the forefront of climate denial, manufacturing doubt about the scientific consensus that its own scientists had confirmed." The reporting wasn't just hurling accusations, Rosero saw. Much of it quoted internal Exxon documents.

Rosero read the reporting, and he saw one document as particularly disturbing: that notorious 1998 memo authored by the American Petroleum Institute, Exxon, Chevron, and several other fossil fuel companies, which stated that "Victory will be achieved when... average citizens 'understand' (recognize) uncertainties in climate science."

Rosero's doubts grew more intense when he read a 2017 peer-reviewed paper about Exxon from Harvard researchers Naomi Oreskes and Geoffrey Supran. The paper, the first of its kind ever published in an academic journal, analyzed dozens of advertisements paid for by Exxon, and contrasted those ads with the company's own internal research on climate change. The researchers concluded that the company had for at least a decade made misleading statements "designed to reach and influence the public." Oreskes and Supran cited a full-page *New York Times* ad that ran in March 2000, in which Exxon questioned whether climate change was actually happening and even argued "increased levels of carbon dioxide can promote crop and forest growth."

Exxon reacted by accusing Oreskes and Supran themselves of spreading misinformation. It claimed that the researchers were "misrepresenting our position on climate change and related research to the public" and that the company "is committed and actively working to reduce the risks posed by climate change." But that did nothing to reassure Rosero, who was stunned by the paper's conclusions. He could see plainly that Oreskes and Supran were using actual Exxon documents as evidence for their argument.

Rosero was filled with questions. "I really wanted to talk to people inside of the company about this," he said. "So I printed it out, I had it by my desk, every chance I had I got people to take a look at it, and read it." He didn't get the reaction he was hoping for. Some people were game to talk about the company's well-documented history of spreading misinformation on climate change, but for the most part it was an uncomfortable topic to bring up. Rosero felt torn. He'd spent good years at Exxon. He'd made friendships, worked on intellectually fulfilling projects, and earned a high salary that helped support his wife and daughter. But he needed to know why Exxon kept saying it cared about climate change while doing things that made the emergency worse.

Rosero tried to imagine himself as an executive at the company: "If I knew the world was changing, and if I knew that in ten years, twenty years, or whatever, that what I'm doing right now has some massive impact, why wouldn't I choose the right thing to do?"

Growing increasingly frustrated, Rosero began posting about climate change on internal Exxon message boards. In December 2019, he wrote on one board that if the United States, the European Union, or others took aggressive action to lower emissions, it could make many of the company's investments worthless: "Given a shrinking carbon budget, how much discovered oil and gas do you estimate will stay in the ground and won't be produced? How much of our portfolio is at risk to be impaired?" His fellow employees seemed unsure how to react. "When we consider investing in new acreage, is this a criteria?" someone responded.

Rosero's statements were sometimes received with outright negativity. After he posted a message that was critical of "climate change denial," someone anonymously replied that, "If you believe what 'climate scientists' told you, you should not work for an oil and gas company." It went on, "XOM [Exxon] may be better off without those working against it and not paying them."

Around this time, some Exxon employees received an email inviting them to a "town hall" event on January 8, 2020, where they were encouraged to ask questions of managers and executives. No topic was apparently off limits, whether that was climate change, energy demand, or the company's oil and gas operations. These were all the components making up Exxon's "dual challenge" slogan, which emphasized the need to provide affordable fossil energy for the world while also becoming more sustainable. Rosero saw this as his chance to air his concerns.

He came to the town hall ready to speak. "Thank you for the opportunity to discuss this critical topic," Rosero said when it was his turn to ask a question. He had written down what he was going to say ahead of time, and he stuck to the script. "I think that the documented efforts of the industry lobby promoting obfuscation and [climate] denial were short-sighted and irresponsible, and our sponsorship of it, shameful. Yet, none of us here participated in those decisions." The problem, as Rosero put it to the room, was the company's unwillingness to atone for the denial, to move past it. It still wasn't too late for Exxon to take genuine action to fix the climate emergency. So why wasn't the company doing that? "We acknowledge the need to reduce our emissions, yet they are set to increase by at least 20 percent over the next five years," he said. "In the end, wouldn't you agree that this is a problem of behaviors and leadership—not science and technology?"

He concluded his remarks at the town hall by questioning why Exxon's leaders "seem unable to understand the magnitude of the problem, are ill-prepared for the urgency of the crisis, and want us to take comfort on the cop-out of the 'dual challenge.'"

Rosero recalled that he could see some people nodding silently. But Tolu Ewherido, an Exxon vice president, was apparently not one of them. "I don't hear a question in there," she reportedly responded. The event moved on to other topics. A follow-up email sent later by Exxon management thanked participants for their passion and energy. It was as if Rosero had never spoken at all. But some of his superiors had taken careful notice of his remarks. Not long after, according to Rosero, the backlash against him began.

"There was an implication that I should have understood not to say anything," he said.

Exxon employees can earn high six-figure salaries and work at the company for decades. But nobody can take their status at Exxon for granted. Each year, employees are graded according to an internal rating system, which determines their salary and job security. As a trusted technical expert, Rosero had received consistently high ratings in previous years, putting him among the top third of employees. He confirmed this to reporters by sharing his official performance review documents. That year seemed to be no different. In April 2020 Rosero met with his supervisor, who informed him that he'd been doing great work for the company. Rosero had every reason to expect that he would once again receive a high score.

But when Rosero was informed of his official ranking several months later, he was shocked to see that he'd been downgraded to the lowest score employees can receive. Rosero was given the option of going on a three-month "Performance Improvement Plan," a designation referred to by other employees simply as "PIP." Complying with the plan would mean having regular checkup meetings with his supervisor. It was like being on corporate probation. He could be fired at any moment. If Rosero didn't want to do that, Exxon informed him, he also had the option of resigning. Rosero didn't need much time to think it over. He decided to quit.

Before he left the company where he'd worked for over a decade, Rosero wrote a parting email to his colleagues and supervisors, many of whom he'd become close with over the years. "With relief and optimism, I'm resigning to pursue a different career path," he wrote. Rosero claimed that his low performance rating was punishment for asking a hard question about Exxon's climate change denial during the town hall meeting. "My observation was deemed 'rude' and my 'tone' disrespectful," he wrote. "So, beware: there are things more important than technical contributions, and there is a price to pay for challenging the status quo."

A spokesperson for Exxon disputed this version of events. "We cannot comment on Mr. Rosero's personnel matters during his time with the company," the company's corporate media relations manager, Casey Norton, later said. "At ExxonMobil, we encourage an open dialogue and we do not tolerate retaliation. We reject any claim that annual performance is evaluated on a single event."

Yet others at the company think that's exactly what happened. "I know Enrique Rosero and his work," one person wrote on a message board discussing the town hall incident. "He was on all the high-value projects. He was a talented employee who worked hard in the general interest. He was brave and spoke truth to power—with 'courage of conviction.' And they PIPed him."

After leaving the company, Rosero posted on LinkedIn, where he shared more details of what happened. "I contributed to all major deep water projects in the portfolio. I've championed innovation, developed technology, been recognized as a company expert and enjoyed successes," he wrote. "I believed in change from within and I gave it my best shot." The result, he said, was "modest gains, lots of pushback."

In that post, Rosero said he no longer wanted to work in the oil and gas industry. "I've come to the conclusion that I must shift my professional efforts to an organization actively working to address the challenge of our generation: climate change," he explained. "I'm pursuing positions in market-based climate change solutions, risk mitigation, and renewable energy." Rosero spent the next few months interviewing with companies and eventually landed a job with AIR Worldwide, a risk modeling firm that provides data about climate risks to corporate customers.

Looking back on that whole turbulent year, Rosero has no regrets about leaving Exxon. "I felt relief," he said. "I no longer felt that I was betraying my conscience."

"Is there risk?"

A MONTH AFTER Joe Biden defeated Donald Trump and became president-elect of the United States, Canadian prime minister Justin Trudeau unveiled an aggressive-sounding plan for fighting the climate emergency. A tax on carbon pollution that Trudeau's government had brought in several years earlier would begin rising by $15 per year. By 2030, the carbon price was set to hit $170 per ton of emissions. "The cleaner your economy, the faster and stronger it will grow," Trudeau said. But the reality was less than inspiring. Due to flexible accounting measures in the policy for large polluters, bitumen producers were likely to pay only "a fraction of the posted carbon price," according to analysis from a think tank called the Canadian Institute for Climate Choices.

As the days counted down to Biden's inauguration, Trudeau made his true priorities clear. The incoming Democratic president had stated clearly on the campaign trail that he intended to cancel for good the Keystone XL pipeline, saying "it is tar sands that we don't need [and] that in fact is a very, very high pollutant." That January, Trudeau made a last-minute attempt to convince Biden to reconsider. "We've had a clear and consistent position supporting this project for years," Trudeau said. "Our government is making sure that Canada's views are heard and considered by the incoming administration at the highest levels." Biden was apparently unconvinced. Within hours of being inaugurated, he signed an executive order blocking the project.

It was tempting to see Biden's election win as the start of a bold new era for climate action. While campaigning against Trump he'd declared that "there is no greater challenge facing our country and the world" than global warming. Biden promised to spend trillions of dollars of federal money making the economy green. "We've already waited too long to deal with this climate crisis and we can't wait any longer," he said on day one of his presidency. "We see it with our own eyes, we feel it, we know it in our bones, and it's time to act." But the disinformation machine that had kept the oil sands profitable through decades of efforts to fight climate change was still alive and well. And its key players were going to fight as hard as they could to make sure this time was no different.

Major oil and gas producers knew it was beneficial for them to say publicly that they were serious about fixing climate change. As Biden announced the details of his Build Back Better Act, a sprawling bill that would result in the federal government spending $3.5 trillion on electric vehicle tax credits, renewable energy projects, and green jobs creation programs, along with social policies like extended childcare and free community college, Exxon announced it was eager to help. The oil and gas producer proposed building a $100 billion project that could capture its carbon emissions and bury them underground. "We want to be part of that conversation," Exxon CEO Darren Woods said.

Because carbon capture and storage was an unproven and expensive technology unlikely to be financed by the private sector alone, Woods suggested that a federal price on carbon emissions could help cover the cost. "There's a variety of mechanisms out there, and I think the challenge will be the finding the right one to incentivize something this large," he said. Exxon at that point was a founding member of an organization called the Climate Leadership Council, which was advocating for an economy-wide price on carbon emissions similar to the policy Trudeau was implementing in Canada. It would start at $40 per ton.

But Greenpeace activists posing as corporate headhunters learned what Exxon really thought about a carbon price when they tricked Keith McCoy, a senior lobbyist in Exxon's Washington team, into sharing his views about climate policy on camera. "Nobody is going to propose a tax on all Americans, and the cynical side of me says, yeah we kind of know that—but it gives us a talking point that we can say, well, what is ExxonMobil for? Well, we're for a carbon tax," McCoy said. "Carbon tax is not going to happen." McCoy confirmed that Exxon was trying to hollow out Biden's spending plan so it only focused on building physical infrastructure and not climate restrictions on oil and gas facilities. "If you lower that threshold, you stick to highways and bridges, then a lot of the negative stuff starts to come out," the Exxon lobbyist said.

Exxon reacted to the political scandal that followed media outlets publishing the video footage by firing McCoy. "The past few days have been disappointing for everyone at ExxonMobil and for me personally," Woods said in a statement. He described McCoy's comments as "disturbing and inaccurate" and said they were "entirely inconsistent with our commitment to the environment." Yet by September, Exxon was spending hundreds of thousands of dollars on Facebook ads attacking the corporate tax increases that would help pay for climate investments in the Build Back Better Act. "Tell Congress no tax hikes," one ad declared.

Charles Koch also wanted to present himself as an ally of the Biden administration. In an interview not long after Biden beat Trump, the fossil fuel billionaire admitted that he "screwed up" by using his vast fortune to fund far-right politics. "Some of the politicians that we had helped get elected, I would see them on TV, and they would be talking about policies that were antithetical—against immigration, against criminal justice reform, against a more peaceful foreign policy," he said. "I was horrified." Koch said that he was intent on "finding ways to work with [Biden] to break down the barriers holding people back."

Yet as the Build Back Better Act made its way through the U.S. Senate, an armada of right-wing groups financed and supported by the Koch network began trying to sink the bill. Americans for Prosperity led a seven-figure ad campaign trying to make the legislation look radical by tying it to Vermont senator and former progressive presidential candidate Bernie Sanders. "Tell Congress to stop the Biden-Sanders spending spree," one ad said. Other groups took aim at the bill's climate change provisions, which if passed as Biden had proposed could reduce America's carbon footprint by nearly one gigaton worth of greenhouse gas emissions, the research firm Rhodium Group calculated. That would be the biggest carbon reduction in U.S. history.

This was terrible news to a Koch-backed organization called the Texas Public Policy Foundation, which attacked the act's climate initiatives as "incredibly expensive and harmful anti-energy policies that will cripple our economy, increase our dependence on foreign oil, and increase cost of living, especially for the poorest Americans." All told, over one hundred groups with ties to Koch Industries came out against Build Back Better. "Fighting the reconciliation bill is a top priority for Charles Koch's surrogates," said Connor Gibson, founder of a corporate watchdog group known as Grassrootbeer Investigations. "This is a viable threat for his network, and we can see that all the tentacles of the Kochtopus are out in full force trying to stop it from passing."

As those efforts intensified, Koch Industries marked a major victory for the oil sands. In late September, the Calgary pipeline builder Enbridge announced that its 340-mile Line 3 pipeline project was finally complete, and oil would begin flowing through it shortly. Indigenous land defenders and environmental groups had for years fought the pipeline expansion, which would bring an additional 370,000 barrels per day of heavy oil from Canada into Minnesota. But the Koch Industries subsidiary Flint Hills Resources, which operated the Pine Bend Refinery, the largest single processor of oil sands crude in the United States, had lobbied heavily in favor of Enbridge's Line 3 project. "The Pine Bend refinery relies exclusively on the Enbridge pipeline system to provide the crude oil it needs," it said in a statement.

When Biden traveled in November 2021 to the international climate negotiations in Glasgow, Scotland, he assured the world that "my administration is working overtime to show that our climate commitment is action, not words." Yet the new Line 3 pipeline completed just weeks earlier would carry bitumen whose annual climate footprint could amount to 193 megatons of carbon emissions, which was equal to the atmospheric damage of a dozen large coal-fired plants. Back in Alberta, premier Jason Kenney was thrilled the project was going forward. "It's fantastic news for our province's economic recovery," he said. "Our government will continue to fight for all projects that help bring our oil to market."

And in Washington, the wheels of climate change disinformation remained well lubricated. When Democrats called in the heads of Exxon, Shell, BP, and Chevron to testify about climate denial, "the oil executives were on message and perfectly in tune with their House Republican allies," concluded Lee Fang in the *Intercept*. "They refused to engage substantively on climate and constantly pivoted to the need to provide cheap gasoline, American jobs, and abundant energy sources." Meanwhile, Neil Crabtree, the pipeline worker invited as a witness by Republicans, generated conveniently distracting media coverage. "Big Oil investigation hears from welder who says he lost job on Keystone pipeline three hours after Biden's inauguration," read a headline in the *Daily Mail*.

After decades of burying the science on climate change, distorting the public debate, financing right-wing operatives, and attacking legislation that could limit emissions, the oil sands industry was still relying on the same tricks to protect its profits. And it was still succeeding. The Build Back Better Act was sliced from $3.5 trillion to $1.75 trillion following objections from Joe Manchin, a conservative Democrat senator from the coal state of West Virginia. In addition to personally profiting from his family coal company, which could have been driven out of business by climate regulations, Manchin had been targeted heavily by Americans for Prosperity and other Koch-backed groups, which urged him to oppose the spending plan. That November, a climate provision that would have forced electric utilities to shift toward greener forms of energy was cut due to Manchin's opposition.

"Joe Manchin . . . I talk to his office every week," the Exxon lobbyist McCoy had said on video earlier that year.

With Biden's climate change agenda hanging in the balance, another Exxon lobbyist, Erik Oswald, went to a late 2021 oil and gas conference in New Mexico and briefed a crowd on how his employer was viewing this current era of climate politics. When secret video of his colleague McCoy was made public, Oswald did damage control for Exxon, calling the statements a "mischaracterization" of the company. But Oswald would soon be at the center of a minor scandal of his own after an audio recording of his remarks at the conference was leaked to the media. "The way we think about this is, not as the crusaders for the climate fix," Oswald explained of Exxon's strategy on climate change. "The way I look at it as a scientist is, all I need to think about is, is there risk? Yes, there's risk. Is it a catastrophic inevitable risk? Not to my mind."

Forty years after an Exxon scientist wrote in an internal memo that climate change "will indeed be catastrophic (at least for a substantial fraction of the earth's population)," the company was still trying to convince people the emergency wasn't real.

Epilogue

WHAT IF THE CONSPIRATORS within the oil sands industry had not been so successful in blocking action to stop the climate emergency? What if at various points over the past six decades, leading oil companies had reckoned with their roles in bringing about the destabilization of our atmosphere and had shared their science and powerful voices with those trying to head it off? What if they had used their political influence to push governments in the United States and Canada to synchronize each country's carbon-reducing efforts, rather than encouraging politicians to undercut each other at every possible moment?

Such thoughts filled my mind as I approached the end of researching and writing this book. And so, in late 2021, I got in touch with Joanna Sustento over Zoom wanting to ask a delicate question. It had been nearly four years since we'd first met in Tacloban City. That initial conversation had taken place in a coffee shop that was underwater during Typhoon Haiyan. I'd been nervous to ask Sustento how she rebuilt her life after having everything she considered most important destroyed by climate change. Over large mugs of coffee, she answered my questions about the worst moments of her life with short, devastating anecdotes, a low-key fury, and the occasional flash of wry humor. Now, through our computer screens, I picked up the conversation: If oil and gas producers hadn't spent decades spreading doubt and denial about climate change, I asked, is it possible that the storm that killed her family never would have happened? "Well yeah," she replied. "I would like to think so."

The climate emergency long predicted within the secret research departments of oil companies has finally arrived. In 2021 alone, more than 2.6 million acres of California went up in flames, an unprecedented cold snap in Texas knocked out electricity for millions of people, and the Canadian town of Lytton, B.C., burned to the ground during a record heat wave only to be bombarded several months later by torrential rains that caused $450 million worth of damage across the region. But the impacts didn't have to be this painful and intense. When Bill McKibben wrote his landmark book about global warming, *The End of Nature*, published in 1989, he wasn't hopeful that we'd be able to stop climate change completely. However, he told me recently, "It did not occur to me that we would perform as badly as we performed. I did not know that governments would essentially do nothing for thirty years."

There's a familiar list of reasons that experts cite for why the world allowed greenhouse gas emissions to keep rising so long, even though scientists like James Hansen were making blatantly clear the planetary chaos those emissions were locking in. We in the West are too addicted to our polluting lifestyles. China and India needed fossil fuels to develop. The economic impacts of shifting to greener forms of energy were too damaging. Our self-serving human nature makes the collective global action required of us impossible. Through his decades of writing and activism on climate change, McKibben came to a much simpler explanation: oil and gas producers lied to protect their profits.

They lied about the science being uncertain. They lied about cleaner industries destroying the economy. They lied about climate change being something for which we are all equally responsible. Another path forward was possible. Imagine, McKibben said, that just hours after Hansen gave his congressional testimony in 1988 waking up the American public to the dangers of global warming, the CEO of Exxon went on *CBS Evening News* and said, "'Our scientists are telling us pretty much the same thing. We've got a real problem and we've got to get to work.'" McKibben said, "If that happens, then we avoid this thirty-year, completely pointless debate about whether global warming is real."

Enrique Rosero agreed that his former employer cost us all a huge early opportunity to get the crisis under control. "There's no doubt we lost decades because of their delay, because of their concerted efforts to undermine science," he said. Rosero knew from his years inside Exxon that this outcome wasn't inevitable. The oil and gas producer wasn't just an early expert in climate science, as were many of its corporate competitors; it was also among the first major companies to study solutions like a price on greenhouse gas emissions. It pained him to think how much climate progress could have been achieved if Exxon had used its global political reach starting in the 1990s to ensure that carbon could no longer be pumped for free into the atmosphere. "That would have significantly changed incentives for everything," Rosero said. "It would have been so much easier to address the crisis if we'd started then."

The timeline for deploying wind and solar and electric vehicles and everything else we need to decarbonize the global economy gets moved up by decades. Developing countries have the tools and technologies they need to lift their citizens out of poverty without destabilizing the climate. By 2021, McKibben said, greenhouse emissions might have already peaked and "we'd be headed down the backside." Our descent would not have to be nearly as rapid or wrenching as it will be with the massive emissions cuts now required. "I think that's the part that's sometimes hard for people to understand. Thirty-three years ago when I wrote *The End of Nature*, we had a variety of options that were fairly modest."

Instead, McKibben said, "We did literally the stupidest stuff we could do." Near the very top of that list was tapping one of the biggest oil reserves on the planet. Not only did the Canadian oil sands lock the United States into an especially polluting form of oil, the industry also intentionally destroyed the political will necessary to get climate change under control. Bitumen from Alberta bankrolled the assault on truth led by companies such as Koch Industries and Exxon. "They're the two most aggressive players in this space during the crucial years. They're the ones that take a political consensus that we better do something and turn it into a Republican consensus that we won't do anything," McKibben said. In the sad multidecade history of why governments didn't act as rapidly as they could to get the climate emergency under control, he said, the oil sands "play an absolutely key role."

During my Zoom call with Sustento, she briefly paused to consider what the world might be like today if oil executives had actually taken seriously the climate warnings given by their own scientists. She chose her words slowly and deliberately. "If they acted in a way that's favorable for the planet, for the people, if they diverted their financial and technological capacity into cleaner sources of energy, we wouldn't be experiencing the climate crisis." Sustento's thoughts returned to the storm that forever altered the course of her life. She said: "It is because of their climate denial, the seed of lies that they planted in our society, that we are the ones who are suffering."

Acknowledgments

THE GENESIS FOR THIS BOOK was a short article that I wrote during the disorienting July days of 2020. It was my first summer in New York City, and thanks to climate change, the hottest ever recorded there. My partner, Kara, and I had moved the previous year from Vancouver, trading one of the most expensive cities in Canada for one of the most expensive in the world. Each day that summer thousands of people came out into the streets to support the racial justice uprising ignited by the police murder of George Floyd. When we could, Kara and I joined on our bicycles. The pandemic was still so new that we were afraid to ride the subway.

That was around the time I read about a fascinating lawsuit proceeding in Colorado, described at length in this book, which sought to hold Suncor and Exxon accountable for disregarding warnings about climate change going back to the 1960s. I knew from growing up in Alberta that these are the two dominant players in the Canadian oil sands. The lawsuit, I realized, could be a legal reckoning for the entire industry. Its allegations provided crucial insight into why, after decade upon decade, the United States and Canada have not moved more aggressively to stop climate change.

I pitched a story to Dave Beers, founding editor of the *Tyee*, the formidable independent media outlet in British Columbia where I have contributed climate reporting ever since graduating from journalism school in 2008. Dave was keen on the story, and even though it attracted only a modest number of readers, he encouraged me to develop it into a book proposal.

We sent it to Greystone Books, which a decade earlier had published Andrew Nikiforuk's seminal book *Tar Sands: Dirty Oil and the Future of a Continent*. My book idea was not quite fully formed, but I knew it contained an important story that hadn't yet been properly told. Greystone's owner and founding publisher Rob Sanders, to his credit, agreed. Absent his and Dave's early enthusiasm, you wouldn't be holding this book in your hands.

I started writing the first draft in January 2021. It was the height of the pandemic's second wave in New York, and with little to do outside our apartment in the freezing cold, Kara and I spent the evenings cooking elaborate meals, watching bad TV, and playing with our cat, Yoko. Kara is from Alberta too, an artist and a deep thinker, and our many conversations about the book helped me pay more attention to the good ideas and discard the bad ones. Jody Rogac and Matt Booth, two close Canadian friends in New York, provided feedback and encouragement during pandemic-pod hangouts stretching long into the night. Regular phone calls with my mom, Diana, as well as my brother, Nigel, and his partner, Megan Jones, helped make those long, lonely winter months much less dreary.

I owe a special thanks to the many wonderful editors I've worked with over the years at *VICE News*. The sections on Typhoon Haiyan and Steve Berman's lawsuits grew out of stories originally commissioned by Harry Cheadle, who's no longer at the outlet. Sections about Rupert Murdoch, Suncor's Colorado pollution, and Exxon's pushing out of Enrique Rosero, along with the Big Oil hearing that opens this book, evolved from reporting I did for Natasha Grzincic, who is among the sharpest and most knowledgeable climate editors I've worked with. Many of my VICE stories also benefited from the eagle-eye editing of Josh Visser and Michael Learmonth.

Other important sections of the book—including the chapter on oil lobbyists battling low-carbon fuel standards and the one about

Jason Kenney's visit to Manhattan—first appeared in much different versions as stories on the *Tyee*. I am probably biased in this regard, but the exceptional team I've worked with there for more than a decade regularly produces some of the best climate journalism in Canada, if not anywhere, with a fraction of the resources enjoyed by more mainstream outlets.

There is an additional *Tyee* piece I should mention, which I reported in late 2019, that provided foundational material for this book.

That story was about Imperial Oil, contrasting a former chairman's assertion that human-caused climate change wasn't happening with internal documents showing the company had known and studied the dangers of global temperature rise for decades. The inspiration for that story was a huge cache of historical documents about Imperial Oil made public by researchers with DeSmog, a media outlet specializing in climate disinformation. Brendan DeMelle, DeSmog's executive director, helped me navigate this vast collection of Imperial files, pointing me toward documents that I relied on heavily during the writing of this book.

I again want to thank Dave Beers, who worked with me closely on the manuscript as a conceptual editor, assisting in developing the book's narrative and central arguments. I also owe much to Paula Ayer, my copy editor at Greystone, who offered invaluable structural feedback while making my sentences sing. They helped turn a passable first draft into a much better book than I imagined possible.

Finally, I want to express huge appreciation for the many people who shared their insights and life experiences with me. It was an absolute pleasure meeting Joanna Sustento in person in the Philippines, and I hope her story, as terrible as it is, causes people to take a deeper look at the injustices of climate change. Enrique Rosero and I had several lengthy conversations about his experiences working for Exxon, providing a rare perspective inside the oil and gas giant for which I am grateful. Steve Berman somehow found time in his extremely busy schedule to meet with me in Seattle, and I have a much fuller knowledge of climate litigation as a result. And Bill McKibben, whose morally electrifying writing on the fossil fuel industry has shaped my own reporting and worldview, also provided valuable input.

These are just a few names among countless others who contributed to this book. Apologies to anyone I may have missed.

Brooklyn, 2022

Notes

INTRODUCTION

the Joe Biden administration was trying to pass: Geoff Dembicki, "CEOs Who Called for Climate Action Now Scrambling to Block Climate Action," *Rolling Stone*, September 21, 2021, www.rollingstone.com/politics/politics-news/build-back-better-reconciliation-bill-business-opposition-1229461/.

were being called out for their tactics by America's top lawmakers: Geoff Dembicki, "Exxon Is Desperate to Keep People From Realizing It Lied About Climate," *VICE News*, October 29, 2021, www.vice.com/en/article/xgd8dw/exxon-accused-lying-climate-science-congressional-hearing.

His voice wavered as he explained: "US Oversight Committee Hearing on Big Oil and the Climate Crisis," Yahoo Finance, October 28, 2021, www.youtube.com/watch?v=08M9cdaJtwo.

a self-recorded video of Crabtree inside a vehicle: "Canceling This Keystone Pipeline to Make a Group of People Happy Has Had Real Life Consequences," Americans for Tax Reform, March 3, 2021, www.atr.org/canceling-keystone-pipeline-make-group-people-happy-has-had-real-life-consequences-we-got-people-who.

one of the authors of the climate change denial playbook: "1998 American Petroleum Institute Global Climate Science Communications Team Action Plan." This document and many others referenced in this book are hosted on the site Climate Files, a database on climate disinformation created by the Climate Investigations Center. The Action Plan document is available at www.climatefiles.com/trade-group/american-petroleum-institute/1998-global-climate-science-communications-team-action-plan/.

coauthored an op-ed: Steve Scalise and Grover Norquist, "Why We Need to Stand Against the Radical Left's War on Energy," *Washington Examiner*, May 18, 2021, www.washingtonexaminer.com/opinion/op-eds/why-we-need-to-stand-against-the-radical-lefts-war-on-energy.

Crabtree's remarks were carried by mainstream outlets: Hiroko Tabuchi and Lisa Friedman, "Oil Executives Grilled Over Industry's Role in Climate Disinformation," *New York Times*, October 28, 2021, www.nytimes.com/2021/10/28/climate/oil-executives-house-disinformation-testimony.html.

I: THE FIRST WARNINGS

"Just another storm"

Sustento shared the small space: The section introducing Joanna Sustento's family comes from her speech at the International Fundraising Congress in Holland in 2020, available at www.youtube.com/watch?v=TYkl_03bqhU&t=97s.

It's also a city of low incomes: Joey A. Gabieta, "Poverty Worsens in Eastern Visayas," *Inquirer.net*, March 20, 2015, https://newsinfo.inquirer.net/680120/poverty-worsens-in-eastern-visayas.

The night before the storm was supposed to hit: This section comes from Sustento's podcast conversation with UNSA Vienna, "How Does It Feel When the Climate Crisis Claims Your Home?," December 28, 2020, www.youtube.com/watch?v=G7eL7mokpbU.

The whole house felt like it was vibrating: This description from Sustento comes from the podcast *Unburnable: The People vs. Arctic Oil*, episode 1, "The Storm," https://podcasts.apple.com/us/podcast/01-the-storm/id1288421569?i=1000393392446.

"looking for clues": UNSA Vienna podcast, "How Does It Feel."

Helicopter pilots saw: "Palau Assesses Damage After Super Typhoon Haiyan," ABC News (Australian Broadcasting Corporation), November 7, 2013, www.abc.net.au/news/2013-11-07/an-super-typhoon-hainan-hits-palau/5075198?nw=0&r=Gallery.

"Let us remain calm": "Mass Evacuations Underway as Super Typhoon Haiyan/Yolanda Heads for Philippines," ABC News (Australian Broadcasting Corporation), November 7, 2013, www.abc.net.au/news/2013-11-07/an-philippines-braces-for-super-typhoon-haiyan/5077590?nw=0&r=Gallery.

This water releases some of the heat: "How Is Climate Change Affecting the Philippines?," Climate Reality Project, January 19, 2016, www.climaterealityproject.org/blog/how-climate-change-affecting-philippines.

CNN reported live on the approaching typhoon: "Typhoon Haiyan One of the Biggest Storms Ever," CNN, November 7, 2013, www.youtube.com/watch?v=Tov6Gol83Fo.

"Men on a hunt"

The warning came during an event in New York City: Benjamin Franta, "On Its 100th Birthday in 1959, Edward Teller Warned the Oil Industry About Global Warming," *Guardian*, January 1, 2018, www.theguardian.com/environment/climate-consensus-97-per-cent/2018/jan/01/on-its-hundredth-birthday-in-1959-edward-teller-warned-the-oil-industry-about-global-warming.

the stone steps of Columbia's Low Library: This description of the building is taken from the Wikipedia entry "Low Memorial Library," last updated on December 25, 2021, https://en.wikipedia.org/wiki/Low_Memorial_Library.

"It pleases me to note that on its hundredth birthday": This quote from Dunlop and subsequent references to his speech are taken from Columbia University, *Energy and Man: A Symposium* (New York: Appleton-Century-Crofts, Inc., 1960), https://books.google.com/books?id=NRpJAAAAMAAJ, 26.

"Endowed with remarkable recall": "He Made Sun Oil Rise: Robert G. Dunlop, W'31, Hon'72," *Wharton Magazine*, July 1, 2007, https://magazine.wharton.upenn.edu/issues/anniversary-issue/he-made-sun-oil-rise-robert-g-dunlop-w31-hon72/.

A black-and-white film produced by the institute: This reference is taken from a 1950s educational film called *Barrel Number One*, produced by the American Petroleum Institute and available at www.youtube.com/watch?v=tn1BiY_bKoo.

Teller began his talk at Columbia: All references to Teller's speech are from Columbia University, *Energy and Man*, 52.

an article in Time *magazine*: Matt Smith, "The Oil Industry Was Warned About Climate Change in 1968," *VICE News*, April 15, 2016, www.vice.com/en/article/mbn4eb/the-oil-industry-was-warned-about-climate-change-in-1968.

This included Project Gabriel: Jill Lepore, "The Atomic Origins of Climate Science," *New Yorker*, January 30, 2017, www.newyorker.com/magazine/2017/01/30/the-atomic-origins-of-climate-science.

Royal Dutch Shell scientist Dr. M. A. Matthews published a paper: Dr. M. A. Matthews, "The Earth's Carbon Cycle," *New Scientist*, October 1959, www.climatefiles.com/shell/1959-shell-earths-carbon-cycle-article/.

"A gift from God"

It had rained the night before Dunlop arrived: Hereward Longley, "Digging for Oil in the Boreal Forest," Hagley Museum and

Library, March 18, 2019, www.hagley.org/librarynews/digging-oil-boreal-forest.

an official at the Hudson's Bay Company heard a report: "Thanadelthur," Alberta Culture and Tourism, http://history.alberta.ca/energyheritage/oil/pre-modern-global-history/early-human-history-in-canada/thanadelthur.aspx.

a British geologist named Dr. T. O. Bosworth traveled to the region: Graham Taylor, *Imperial Standard: Imperial Oil, Exxon, and the Canadian Oil Industry From 1880* (Calgary: University of Calgary Press, 2019), https://prism.ucalgary.ca/bitstream/handle/1880/110195/9781773850368_chapter10.pdf, 215.

Pew was not known for his warmth or charisma: Darren Dochuk, "The Other Brother Duo That Brought Us the Modern GOP," *Politico*, September 2, 2019, www.politico.com/magazine/story/2019/09/02/pew-brothers-politics-influence-wealth-227993/.

He saw the oil sands as a potential source of new profits: "J. Howard Pew," Alberta Culture and Tourism, http://history.alberta.ca/energyheritage/sands/mega-projects/experimentation-and-commercial-development/j-howard-pew.aspx.

The oil executive was a Presbyterian fundamentalist: Editorial, "J. Howard Pew, 1882–1971," *Christianity Today*, December 17, 1971, www.christianitytoday.com/ct/1971/december-17/editorials-j-howard-pew-18821971.html.

Pew's religious convictions would end up aiding Sun Oil: Darren Dochuk, "Prairie Fire: The New Evangelicalism and the Politics of Oil, Money, and Moral Geography," in *American Evangelicals and the 1960s*, ed. Axel R. Schäfer (Madison, WI: University of Wisconsin Press, 2013), 39–60.

But the premier's religious beliefs: "Nuking the Oilsands: Why Ernest Manning Wanted Nuclear Weapons to Jump-start Alberta's Oil Industry," CBC Radio, *The Current*, July 17, 2019, www.cbc.ca/radio/thecurrent/the-current-for-july-17-2019-1.5214680/nuking-the-oilsands-why-ernest-manning-wanted-nuclear-weapons-to-jumpstart-alberta-s-oil-industry-1.5214690.

Pew's personal obsession with the oil sands: "Downstream Impact: Building the Canadian Tar Sands Industry With Hereward Longley," Hagley Museum and Library, www.hagley.org/downstream-impact-building-canadian-tar-sands-industry-hereward-longley.

"Buckets broke teeth on the frozen ground": "Industry Landmark: The Great Canadian Oil Sands Plant," Alberta Culture and Tourism, www.history.alberta.ca/energyheritage/sands/mega-projects/experimentation-and-commercial-development/industry-landmark-the-great-canadian-oil-sands-plant.aspx.

But that didn't stop dozens of airplanes and jets from arriving: Reid Southwick, "Oilsands @ 50: Triumph Over Challenges Gave Rise to Alberta's Oilsands," *Calgary Herald*, September 25, 2017, https://calgaryherald.com/business/energy/oilsands-50-triumph-over-challenges-gave-rise-to-albertas-oilsands.

Dochuk later said of the moment: "Nuking the Oilsands," *The Current*.

"We were all so desperate"

"We were all looking outside": All quotes from Joanna Sustento in this chapter come from the autobiographical article "The Impossible—A Story About Love, Decisions and Survival. A Story About Haiyan," Storya.ph, July 30, 2014, https://web.archive.org/web/20190812195318/http://www.storya.ph/joanna-sustento-story-53ce07e6b9769.

A "hellish cloud"

A worker named George Skulsky remembers a night: Dan Healing, "Oilsands Pioneers Recall Big Promise, Big Problems With Industry's First Mine," CBC, June 21, 2017, www.cbc.ca/news/canada/edmonton/oilsands-history-pioneers-industry-oil-alberta-1.4171770.

On one day in 1943, the smog got so bad: Jess McNally, "July 26, 1943: L.A. Gets First Big Smog," *Wired*, July 26, 2010, www.wired.com/2010/07/0726la-first-big-smog/.

Its mission was to fund and conduct industry research: Charles A. Jones, "A Review of the Air Pollution Research Program of the Smoke and Fumes Committee of the American Petroleum Institute," *Journal of the Air Pollution Control Association* 8, no. 3 (1958), doi: 10.1080/00966665.1958.10467854.

as the story of Harold Johnston makes clear: All references to Johnston's story come from a 1999 interview with Sally Smith Hughes as part of the Regional Oral History Office at the University of California, Berkeley, https://digitalassets.lib.berkeley.edu/roho/ucb/text/johnston_harold.pdf.

Years later, the Smoke and Fumes Committee was still arguing: Vance N. Jenkins, "The Petroleum Industry Sponsors Air Pollution Research," *Air Repair* 3, no. 3 (1954), www.smokeandfumes.org/documents/4.

recognized at the highest levels of the U.S. government: Dana Nuccitelli, "Scientists Warned the US President About Global Warming 50 Years Ago Today," *Guardian*, November 5, 2015, www.theguardian.com/environment/climate-consensus-97-per-cent/2015/nov/05/scientists-warned-the-president-about-global-warming-50-years-ago-today.

The scientists warned the Democratic president: President's Science Advisory Committee, "Restoring the Quality of Our Environment," White House, November 1965, https://carnegiedge.s3.amazonaws.com/downloads/caldeira/PSAC,%201965,%20Restoring%20the%20Quality%20of%20Our%20Environment.pdf.

The oil industry was not pleased: The oil industry's reaction to the Johnson report is contained in a document uncovered by researcher Benjamin Franta, entitled "Proceedings 1965," available at www.climatefiles.com/trade-group/american-petroleum-institute/1965-api-president-meeting-the-challenges-of-1966/.

The duo delivered their report in 1968: Center for International Environmental Law, "Smoke and Fumes: The Legal and Evidentiary Basis for Holding Big Oil Accountable for the Climate Crisis," November 2017, www.ciel.org/wp-content/uploads/2019/01/Smoke-Fumes.pdf.

Sun Oil began sending the very first oil: "The Oil Sands Story (1960s, 1970s & 1980s)," Suncor, www.suncor.com/en-ca/who-we-are/history/the-oil-sands-story.

"Operation Oilsands"

The first iteration of Imperial Oil was Canadian: "Our History," Imperial Oil, last updated 2021, www.imperialoil.ca/en-CA/Company/About/Our-history.

the company concluded that actually transforming the thick bitumen into fuel: Graham Taylor, *Imperial Standard: Imperial Oil, Exxon, and the Canadian Oil Industry From 1880* (Calgary: University of Calgary Press, 2019), https://prism.ucalgary.ca/bitstream/handle/1880/110195/9781773850368_chapter10.pdf, 215.

One of the project's most enthusiastic backers was Edward Teller: "Plowshare Program," U.S. Office of Scientific and Technical Information, www.osti.gov/opennet/reports/plowshar.pdf.

The program also piqued the interest of Manley Natland: David Breen, *Alberta's Petroleum Industry and the Conservation Board* (Edmonton: University of Alberta Press, 1993), https://archive.org/details/albertaspetroleu0000bree/page/452/mode/2up?q=project+cauldron, 440.

A front-page story in the Calgary Herald: Taylor, *Imperial Standard*, 221.

A health official named Dr. D. Dick was also worried: Breen, *Alberta's Petroleum Industry and the Conservation Board*, 519.

By 1968, the projected costs were $800 million: Taylor, *Imperial Standard*, 225.

The warning was contained in a 1970 report: H. R. Holland, "Pollution Is Everybody's Business," Imperial Oil, January 30, 1970, www.climatefiles.com/exxonmobil/1970-exxon-imperial-oil-pollution-is-everybodys-business/.

it was unclear if Imperial Oil's oil sands project at Mildred Lake: Taylor, *Imperial Standard*, 226.

"The dream was saved and construction hit fever pitch": "Our History," Syncrude, last updated 2022, https://syncrude.ca/our-project/our-history/.

During a presentation at corporate headquarters: J. F. Black, "The Greenhouse

Effect," Exxon Research and Engineering Company, June 6, 1978, https://insideclimatenews.org/wp-content/uploads/2015/09/James-Black-1977-Presentation.pdf.

There were six hundred guests there that day: Suzanne Zwarun, "Getting Into the Flow," *Maclean's*, September 25, 1978, https://archive.macleans.ca/article/1978/9/25/getting-into-the-flow.

"A million pieces"

As Joanna Sustento and her mother fought the pull: All quotes from Joanna Sustento in this chapter are from Joanna Sustento, "The Impossible—A Story About Love, Decisions and Survival. A Story About Haiyan," Storya.ph, July 30, 2014, https://web.archive.org/web/20190812195318/http://www.storya.ph/joanna-sustento-story-53ce07e6b9769.

Another woman was swept into the waves with her five-year-old son: "Typhoon Haiyan: Survivors' Stories in the Philippines," BBC, November 11, 2013, www.bbc.com/news/world-asia-24896060.

Another man lost twenty family members: Charity Durano and Gail DeGeorge, "Three Years After Super Typhoon, Survivors Share Stories of Rebuilding," *National Catholic Reporter*, October 5, 2016, www.ncronline.org/news/world/three-years-after-super-typhoon-survivors-share-stories-rebuilding.

II: THE EARLY CONSTRUCTION OF DENIAL

"He seemed embarrassed"

As Charles Koch flew over the Alaskan wilderness: Christopher Leonard, *Kochland: The Secret History of Koch Industries and Corporate Power in America* (New York: Simon & Schuster, 2019), 44.

But he also had the publicity-avoiding temperament of someone: Jane Mayer, *Dark Money: The Hidden History of the Billionaires Behind the Rise of the Radical Right* (New York: Anchor Books, 2017), 51.

Among the assets that Charles and his brothers had inherited: Leonard, *Kochland*, 50.

The company would later refer to these oil sources: This reference was taken from legal documents from a lawsuit that William Koch filed against Koch Industries in 1997, available at https://casetext.com/case/koch-v-koch-industries-inc-4.

that argued during the 1960s that the civil rights movement: Gladwin Hill, "Birch Head Sees Red Rights Plot," *New York Times*, August 16, 1963, www.nytimes.com/1963/08/16/archives/birch-head-sees-red-rights-plot-welch-calls-aim-formation-of-soviet.html.

Charles apparently rolled his eyes: Mayer, *Dark Money*, 52.

The new think tank's ideology could be summed up: Charles Koch, "The Case for a Free Market in Energy," *Libertarian Review*, August 1977, https://kochdocs.org/2019/07/27/charles-koch-case-free-market-energy/.

J. Howard Pew was a staunch anti-communist: Darren Dochuk, "The Other Brother Duo That Brought Us the Modern GOP," *Politico*, September 2, 2019, www.politico.com/magazine/story/2019/09/02/pew-brothers-politics-influence-wealth-227993/.

Starting in the late 1960s, Koch Industries began: David Sassoon, "Koch Brothers' Political Activism Protects Their 50-Year Stake in Canadian Heavy Oils," *Inside Climate News*, May 10, 2012, https://insideclimatenews.org/news/10052012/koch-industries-brothers-tar-sands-bitumen-heavy-oil-flint-pipelines-refinery-alberta-canada/.

the early bitumen they processed was far outside the mainstream: William Koch lawsuit against Koch Industries, https://casetext.com/case/koch-v-koch-industries-inc-4.

Charles's decision to gamble on obscure sources of oil: Leonard, *Kochland*, 86.

"Ahead of the game"

Celina Harpe of the Fort McKay First Nation: Ted Genoways, "The High Cost of Oil," *Outside*, November 11, 2014, www.outsideonline.com/outdoor-adventure/environment/high-cost-oil/.

the Fort McKay chief took the oil sands producer to court: Brandi Morin, "Fort McKay, a First Nation That Survived the Past, and Is Looking Towards the Future Part 1," *APTN News*, June 22, 2015, www.aptnnews.ca/

national-news/fort-mckay-first-nation-survived-past-looking-towards-future-part-1/.

The document warned that one result of this ecological anxiety: "1973 Imperial Oil Review of Environmental Protection Activities," Imperial Oil, 1973, www.climatefiles.com/exxonmobil/1973-imperial-oil-review-of-environmental-protection/.

Another internal document from this time period made the connection explicit: Public Affairs Department, "Canadian Pressure Groups, Part 1," Imperial Oil, May 1976, www.documentcloud.org/documents/5015298-1976-Canadian-Pressure-Groups-by-Public-Affairs.html.

Imperial Oil had begun to realize that public concern: Frank T. Lebart, "Air/Water Pollution in Canada: A Public Relations Assessment for Imperial Oil Limited," Imperial Oil, 1967, www.climatefiles.com/exxonmobil/1967-imperial-oil-public-relations-assessment-of-air-water-pollution-in-canada/.

Imperial Oil began actively monitoring "public pressure groups": Public Affairs Department, "Canadian Pressure Groups, Part 1."

Exxon predicted that concerns about air and water pollution were just the start: This reference comes from a 1979 memo that Exxon's Henry Shaw wrote to H. N. Weinberg about research on atmospheric science, available at www.climatefiles.com/exxonmobil/1979-exxon-memo-on-atmospheric-science-research-to-influence-legislation/.

The threat posed by such groups was real: "Review of Environmental Protection Activities for 1978–1979," Imperial Oil, 1980, www.climatefiles.com/exxonmobil/1980-imperial-oil-review-of-environmental-protection-activities-for-1978-1979/.

Exxon made studying climate science a top priority: Neela Banerjee, Lisa Song, and David Hasemyer, "Exxon's Own Research Confirmed Fossil Fuels' Role in Global Warming Decades Ago," *Inside Climate News*, September 16, 2015, https://insideclimatenews.org/news/16092015/exxons-own-research-confirmed-fossil-fuels-role-in-global-warming/.

A senior Exxon scientist would later tell the media outlet: Lisa Song, Neela Banerjee, and David Hasemyer, "Exxon Confirmed Global Warming Consensus in 1982 With In-House Climate Models," *Inside Climate News*, September 22, 2015, https://insideclimatenews.org/news/22092015/exxon-confirmed-global-warming-consensus-in-1982-with-in-house-climate-models/.

Imperial Oil was also aware of this climate research: "Review of Environmental Protection Activities for 1978–1979," Imperial Oil.

Just how severe was sketched out during a 1980 meeting: This reference is taken from a document showing minutes of a 1980 American Petroleum Institute meeting with its CO_2 and Climate Task Force, available at www.climatefiles.com/climate-change-evidence/1980-api-climate-task-force-co2-problem/.

Within Exxon there continued to be scientific debates: This reference is taken from a 1981 memo that Exxon's Roger Cohen wrote to Exxon's Werner Glass, available at www.climatefiles.com/exxonmobil/1981-exxon-memo-on-possible-emission-consequences-of-fossil-fuel-consumption/.

From 1983 to 1988, its Syncrude consortium spent $1.6 billion: Marc Humphries, "North American Oil Sands: History of Development, Prospects for the Future," Congressional Research Service, January 17, 2008, www.everycrsreport.com/files/20080117_RL34258_6f7175d6e33a0875b3b732cf40311f600edc4720.pdf.

The 1980s were a tough period for Suncor: "The Oil Sands Story (1960s, 1970s & 1980s)," Suncor, last updated 2022, www.suncor.com/en-ca/who-we-are/history/the-oil-sands-story.

Ken Croasdale, a Calgary-based scientist who researched global heating: Sara Jerving, Katie Jennings, Masako Melissa Hirsch, and Susanne Rust, "What Exxon Knew About the Earth's Melting Arctic," *Los Angeles Times*, October 9, 2015, https://graphics.latimes.com/exxon-arctic/.

"Very strong interests at stake"

Rivera interviewed railroad employees: Shell Chemical Company, "Emerging Techniques for Effective Corporate Response to Public Issues," *TREND*, 1980, www.climatefiles.com/shell/1980-shell-chemical-company-trend-publication/.

the company commissioned the University of East Anglia: All references to Shell's commissioning of the University of East Anglia to study climate change are taken from Appendix 8 of a "confidential" report from the company uncovered by Jelmer Mommers of *De Correspondent*, available at www.climatefiles.com/shell/1988-shell-report-greenhouse/.

the European oil giant was leading a $13 billion proposal: "Shell May Quit Alsands," *New York Times*, March 17, 1982, www.nytimes.com/1982/03/17/business/shell-may-quit-alsands.html.

Shell eventually decided to scale back: Reuters, "Shell Canada Plant," *New York Times*, May 7, 1982, www.nytimes.com/1982/05/07/business/shell-canada-plant.html.

The "crowded official opening" was covered by the Calgary Herald: Gordon Jaremko, "Refinery Sign of Better Times," *Calgary Herald*, September 27, 1984, https://news.google.com/newspapers?nid=Hx6RvaqUy91C&dat=19840927&printsec=frontpage&hl=en.

Shell knew full well this project would accelerate: From Shell's "confidential" report uncovered by Jelmer Mommers of *De Correspondent*, www.climatefiles.com/shell/1988-shell-report-greenhouse/.

"Pitted against our very survival"

There were two rows of television-camera lights pointed directly at him: Nathaniel Rich, "Losing Earth: The Decade We Almost Stopped Climate Change," *New York Times Magazine*, August 1, 2018, www.nytimes.com/interactive/2018/08/01/magazine/climate-change-losing-earth.html.

His remarks made front-page news the next day: Philip Shabecoff, "Global Warming Has Begun, Expert Tells Senate," *New York Times*, June 24, 1988, www.nytimes.com/1988/06/24/us/global-warming-has-begun-expert-tells-senate.html.

then it would be less risky for politicians to take action: Rich, "Losing Earth."

policymakers from dozens of countries gathered in Toronto: Conference Statement, "The Changing Atmosphere: Implications for Global Security," Toronto, 1988, www.academia.edu/4043227/The_Changing_Atmosphere_Implications_for_Global_Security_Conference_Statement_1988.

more than two dozen climate change bills working through Congress: Rich, "Losing Earth."

the CEO of Imperial Oil, Arden Haynes, gave a speech: Arden Haynes, "Canada and Its Energy: Opportunities Waiting to Happen," Empire Club of Canada, January 16, 1989, www.climatefiles.com/exxonmobil/1989-imperial-oil-speech-canada-and-its-energy-opportunities-waiting-to-happen/.

a management committee put together a report for leaders of Esso: G. W. Schindel, "Brundtland Report Overview/Implications," Esso Resources Canada Limited, February 16, 1988, www.climatefiles.com/exxonmobil/1988-esso-presentation-on-brundtland-report-implications/.

III: SOLUTIONS KNOWN AND SABOTAGED

"Threaten the existence"

In 1991, Imperial Oil figured out a way to stop climate change: Imperial Oil, "A Discussion Paper on Global Warming Response Options," April 1991, www.climatefiles.com/exxonmobil/imperial-oil/1991-imperial-oil-global-warming-response-options/.

This was part of a $3 billion "Green Plan": Minister of Supply and Services Canada, "Canada's Green Plan In Brief," Government of Canada, 1990, www.gulfofmaine.org/resources/gomc-library/CA%20green%20plan%20in%20brief.pdf.

"the average cost of producing a barrel of light crude": Arden Haynes, "Canada and Its Energy: Opportunities Waiting to Happen," Empire Club of Canada, January 16, 1989, www.climatefiles.com/exxonmobil/1989-imperial-oil-speech-canada-and-its-energy-opportunities-waiting-to-happen/.

And that in turn resulted in higher greenhouse gas emissions: Imperial Oil, "A Discussion Paper on Global Warming Response Options."

Imperial expected this trend to continue: Esso Resources Canada Ltd., "The Potential for CO_2 Reductions From Additional Energy Efficiency," May 1991, www.climatefiles.com/exxonmobil/1991-imperial-report-potential-for-co2-reductions-from-energy-efficiency/.

The same went for capturing the carbon released from its bitumen plants: Imperial Oil, "Underground Disposal of Carbon Dioxide," April 1991, www.climatefiles.com/exxonmobil/1991-imperial-oil-discussion-paper-on-underground-disposal-of-carbon-dioxide/.

Canada's emissions would plateau around 1990 levels and then begin to shrink: Imperial Oil, "A Discussion Paper on Global Warming Response Options."

places that were reliant on fossil fuels could be hammered: James Osten, George Vasic, and David West, "Carbon Dioxide Emissions and Federal Energy Policy," DRI/McGraw-Hill, March 18, 1991, www.documentcloud.org/documents/5015325-Carbon-Dioxide-Emissions-and-Federal-Energy.html.

The oil sands were especially vulnerable: Imperial Oil Limited, "Detailed Issue Summary: Global Warming/Climate Change," November 29, 1993, www.climatefiles.com/exxonmobil/1993-imperial-oil-issue-summary-of-global-warming-climate-change/.

Imperial's strategy was laid out in a 1993 document: Imperial Oil Limited, "Detailed Issue Summary: Global Warming/Climate Change."

This was a misrepresentation of Imperial's own research: Osten, Vasic, and West, "Carbon Dioxide Emissions and Federal Energy Policy."

"I feel embarrassed"

Meeting at the Capital Hilton hotel: All references in this chapter to the Cato Institute denier event are taken from a flyer for the 1991 conference entitled "Global Environmental Crisis: Science or Politics" in the Koch Docs online archive, https://kochdocs.org/2019/08/12/1991-cato-climate-denial-conference-flyer-and-schedule/.

proceeded to give it as much as $20 million in start-up funding: Jane Mayer, *Dark Money: The Hidden History of the Billionaires Behind the Rise of the Radical Right* (New York: Anchor Books, 2017), 107.

A Goldman Sachs analysis from the 1980s estimated the value: This reference was taken from legal documents from a lawsuit that William Koch filed against Koch Industries in 1997, available at https://casetext.com/case/koch-v-koch-industries-inc-4.

this new source of heavy oil sands crude could be shipped via pipeline: David Sassoon, "Koch Brothers' Political Activism Protects Their 50-Year Stake in Canadian Heavy Oils," *Inside Climate News*, May 10, 2012, https://insideclimatenews.org/news/10052012/koch-industries-brothers-tar-sands-bitumen-heavy-oil-flint-pipelines-refinery-alberta-canada/.

That campaign was organized by Southern Energy: References to the 1991 climate denial campaign run by Southern Energy and others are taken from a collection of Information Council for the Environment documents hosted on the Climate Files archive, available at www.climatefiles.com/denial-groups/ice-ad-campaign/.

"We can look down the road a little way and see an industry under siege": Mayer, *Dark Money*, 248.

"We have to get this right"

EDF's plan for stopping climate change had two key components: Eric Pooley, *The Climate War: True Believers, Power Brokers, and the Fight to Save the Earth* (New York: Hachette Books, 2010), 57.

"The time for study has ended," Bush said: David Espo, "Bush Pledges to Reduce Acid Rain," Associated Press, August 31, 1988, https://apnews.com/article/18545a789531a726fb538ecce93a8658.

Once Bush was in office, his advisers requested a meeting with EDF's Fred Krupp: Pooley, *The Climate War*, 177.

the acid rain legislation led to an estimated $119 billion in net benefits annually: John Whitehead, "Benefits and Costs of the Acid Rain Program," *Environmental Economics* (blog), October 31,

2005, www.env-econ.net/2005/10/benefits_and_co.html.

it contained only rough guidelines for how to do so and no firm timetable: Stephanie Meakin, "The Rio Earth Summit: Summary of the United Nations Conference on Environment and Development," Government of Canada, November 1992, https://publications.gc.ca/Collection-R/LoPBdP/BP/bp317-e.htm.

while giving more flexibility to developing countries: Information Unit for Conventions, "United Nations Framework Convention on Climate Change," United Nations Environment Programme, October 1997, https://unfccc.int/cop3/fccc/info/backgrod.htm.

"Americans can't hear the whistle"

The scientists were brought in to explain the mounting threats: All references to the 1997 Clinton meeting are taken from a document summarizing a meeting that Lenny Bernstein attended on behalf of the Global Climate Coalition, available on the Climate Files archive at www.climatefiles.com/denial-groups/global-climate-coalition-collection/1997-white-house-meeting-and-stac-conference-call/.

a series of ads paid for by Bernstein's organization: Copies of these advertisements are hosted on the Climate Files archive, available at www.climatefiles.com/denial-groups/global-climate-coalition-collection/1997-anti-kyoto-ads/.

The Global Climate Coalition was formed in 1989: "Global Climate Coalition Membership," November 16, 1989, https://www.climatefiles.com/denial-groups/global-climate-coalition-collection/1989-membership/.

the organization wrote to U.S. Republican senator John Heinz: J. Robert Minter, letter addressed to staffer of Senator John Heinz, December 12, 1989, www.climatefiles.com/denial-groups/global-climate-coalition-collection/1989-solicit-senator-meetings/.

An early bulletin produced by the coalition: John Shlaes, "Charting the Future of Global Climate Change," *Climate Watch: The Bulletin of the Global Climate Coalition* 1, no. 1 (December 1992), www.climatefiles.com/denial-groups/global-climate-coalition-collection/1992-climate-watch-bulletin-vol-1-issue-1/.

The Global Climate Coalition was thrilled: "Post Reporter Finds Threat Uncertain," *Climate Watch: The Bulletin of the Global Climate Coalition* 1, no. 9 (September 1993), www.climatefiles.com/denial-groups/global-climate-coalition-collection/1993-climate-watch-vol-1-issue-9/.

Privately, the coalition knew that Michaels: L. S. Bernstein, "Global Climate Coalition (GCC)—Primer on Climate Change Science—Final Draft," Global Climate Coalition, December 21, 1995, www.climatefiles.com/denial-groups/global-climate-coalition-collection/global-climate-coalition-draft-primer/.

"no credible scientific evidence exists which shows that these changes": "What the Media Hasn't Told You About the IPCC Report," *Climate Watch: The Bulletin of the Global Climate Coalition* 3, no. 5 (fourth quarter 1995), www.climatefiles.com/denial-groups/global-climate-coalition-collection/1995-year-end-bulletin-vol-3-issue-5/.

who in his floor speech referenced the dubious science of Patrick Michaels: Eric Pooley, *The Climate War: True Believers, Power Brokers, and the Fight to Save the Earth* (New York: Hachette Books, 2010), 90.

A later study by Environment Canada found: James Armstrong, "Global Benefits and Costs of the Montreal Protocol," in *Protecting the Ozone Layer*, eds. Philippe G. Le Prestre, John D. Reid, and E. Thomas Morehouse Jr. (Boston: Springer, 1998), 173–77.

Imperial Oil commissioned one of the first-ever studies: Imperial Oil, "A Discussion Paper on Global Warming Response Options," April 1991, www.climatefiles.com/exxonmobil/imperial-oil/1991-imperial-oil-global-warming-response-options/.

Imperial's early carbon-tax study was specifically referenced: "1991 Imperial Oil Discussion Paper on Global Warming Response Options," Climate Files, last updated 2022, www.climatefiles.com/exxonmobil/imperial-oil/1991-imperial-oil-global-warming-response-options/.

a 1996 background paper released by the coalition: Global Climate Coalition, "Economic and Lifestyle Impacts From Proposed Greenhouse Gas Emissions Restrictions," 1996, www.climatefiles.com/denial-groups/global-climate-coalition-collection/1996-economic-impacts-from-proposed-emission-restrictions/.

ideas innovated in Alberta's oil patch were amplified south of the border: Climate Investigations Center, "Imperial Oil Document Trove," December 3, 2019, https://climateinvestigations.org/imperial-oil-document-trove/.

the coalition blitzed mainstream media outlets across America: Copies of these advertisements are hosted on the Climate Files archive, at www.climatefiles.com/denial-groups/global-climate-coalition-collection/1997-anti-kyoto-ads/.

"Let's face it: the science of climate change is too uncertain": Amy Lieberman and Susanne Rust, "Big Oil Braced for Global Warming While It Fought Regulations," *Los Angeles Times*, December 31, 2015, https://graphics.latimes.com/oil-operations/.

A major study of global opinion on climate change: Stuart Capstick, Lorraine Whitmarsh, Wouter Poortinga, Nick Pidgeon, and Paul Upham, "International Trends in Public Perceptions of Climate Change Over the Past Quarter Century," *WIREs Climate Change* 6 (2015), doi:10.1002/wcc.321.

President Bush explained in a 2001 letter: President George W. Bush, "Text of a Letter From the President to Senators Hagel, Helms, Craig, and Roberts," White House, March 13, 2001, https://georgewbush-whitehouse.archives.gov/news/releases/2001/03/20010314.html.

A State Department memo from 2001: This reference is taken from unclassified briefing notes for a Bush administration official for her upcoming meeting with members of the Global Climate Coalition, available at www.climatefiles.com/denial-groups/global-climate-coalition-collection/2001-state-department-meeting/.

"The dumbest-assed thing"

Canadian journalists received an invitation to an unusual press event: Press Release, "Kyoto's Fatal Flaws Revealed," APCO Worldwide (Canada), November 12, 2002, https://archive.is/YyfPX.

he'd recently claimed in an article trashing the Kyoto Protocol: Patrick Michaels, "Kyoto Is Useless—and That Is the Best That You Can Say About It," *Free Lance-Star*, June 1, 2002, https://fredericksburg.com/columns/kyoto-is-useless--and-that-is-the-best-that-you-can-say-about-it/article_eb499622-c2b7-5543-92b2-9f446c5ec9d8.html.

Members of Parliament were set to vote the following month: Steven Chase, "Foes of Kyoto Protocol Aim to Stir Up Doubts," *Globe and Mail*, November 13, 2002, www.theglobeandmail.com/news/national/foes-of-kyoto-protocol-aim-to-stir-up-doubts/article25425577/.

Peterson insisted during the lead-up to the Ottawa event: Brent Jang, "Imperial Oil Executives Mince No Words on Kyoto or Anything Else," *Globe and Mail*, March 13, 2002, www.theglobeandmail.com/report-on-business/imperial-oil-executives-mince-no-words-on-kyoto-or-anything-else/article753582/.

They saw him as a person of strong morals: "Obituary: Robert Byron Peterson," Dignity Memorial, www.dignitymemorial.com/obituaries/aurora-on/robert-peterson-10014944.

He got a job as a summer student at Imperial in 1958: "Robert B. Peterson," Canadian Petroleum Hall of Fame, last updated 2019, www.canadianpetroleumhalloffame.ca/robert-peterson.html.

Peterson was suspicious from the start: This reference is taken from a 1995 letter Robert Peterson wrote to then Canadian prime minister Jean Chrétien, available at www.climatefiles.com/exxonmobil/imperial-oil/1995-imperial-oil-letter-from-robert-peterson-to-canadian-government-re-unfccc/.

Peterson was openly attacking the science behind climate change: Imperial Oil, "Annual Report to Shareholders 1996," 1997, www.documentcloud.org/documents/5015211-ar1996.html.

With that statement Peterson was contradicting: Sara Jerving, Katie Jennings, Masako Melissa Hirsch, and Susanne Rust, "What Exxon Knew About the Earth's Melting Arctic," *Los Angeles Times*, October 9, 2015, https://graphics.latimes.com/exxon-arctic/.

Peterson framed his opposition to climate policy in noble-sounding terms: Imperial Oil, "Annual Report to Shareholders 1997," 1998, www.documentcloud.org/documents/5015212-ar1997.html.

to grow from 73,000 barrels to more than 130,000 barrels per day: Imperial Oil, "Annual Report to Shareholders 1996."

Canada's greenhouse gas emissions had soared: Matthew Bramley, "Greenhouse Gas Emissions From Industrial Companies in Canada: 1998," Pembina Institute, October 2000, www.jstor.org/stable/resrep00216?seq=1#metadata_info_tab_contents.

so Peterson lied instead: Robert Peterson, "A Cleaner Canada," *Imperial Oil Review* 26 (Summer 1998), www.climatefiles.com/exxonmobil/imperial-oil/1998-imperial-oil-article-a-cleaner-canada-by-robert-peterson/.

Peterson was especially outspoken among oil sands CEOs: Jang, "Imperial Oil Executives Mince No Words on Kyoto."

meeting the protocol's targets would be no problem: Bernard Simon, "Canada's Oil Sector Fights Pollution Plan," *New York Times*, November 26, 2002, www.nytimes.com/2002/11/26/business/canada-s-oil-sector-fights-pollution-plan.html.

meeting the climate goals of Kyoto would cause $30 billion worth of damage: "Major Canadian Business Groups Take Aim at Kyoto," Institute for Agriculture and Trade Policy, March 6, 2002, www.iatp.org/news/major-canadian-business-groups-take-aim-at-kyoto.

Those two organizations were meanwhile spending: Zoe Cormier, "Coming Clean on Climate-Change Spin—How the PR Industry Sold the "Made in Canada" Solution to Global Warming," *This Magazine*, September–October 2006, https://archive.fo/jLjB7#selection-343.0-343.108.

The oil sands industry was thrilled: Patrick Brethour and Steven Chase, "Kyoto Impact Minimal, Suncor says," *Globe and Mail*, January 10, 2003, www.theglobeandmail.com/news/national/kyoto-impact-minimal-suncor-says/article1009648/.

IV: A PUBLIC AWAKENING

"Victory will be achieved"

gave an unexpected speech at Stanford University: Lester R. Brown, "The Rise and Fall of the Global Climate Coalition," Earth Policy Institute, www.earth-policy.org/mobile/releases/alert6#.

For much of its existence, the Global Climate Coalition: "1989 GCC Membership," Climate Files, last updated 2022, www.climatefiles.com/denial-groups/global-climate-coalition-collection/1989-membership/.

Documents show that the association was also helping Philip Morris: "National Association of Manufacturers," Tobacco Tactics, last updated February 3, 2020, https://tobaccotactics.org/wiki/national-association-of-manufacturers/.

All the strands came together: "The Advancement of Sound Science Coalition (TASSC)," DeSmog, www.desmog.com/advancement-sound-science-coalition/.

"victory will be achieved when": "Global Climate Science Communications—Action Plan," Global Science Team, April 3, 1998, www.climatefiles.com/trade-group/american-petroleum-institute/1998-global-climate-science-communications-team-action-plan/.

"They lied about everything"

For as long as he could recall, Steve Berman disliked the tobacco industry: Geoff Dembicki, "Meet the Lawyer Trying to Make Big Oil Pay for Climate Change," *VICE*, December 22, 2017, www.vice.com/en/article/43qw3j/meet-the-lawyer-trying-to-make-big-oil-pay-for-climate-change.

Berman says it didn't spark his interest: Erik Lundegaard, "Keep It Simple, Stupid," *Super Lawyers*, June 17, 2014, www.superlawyers.com/washington/article/keep-it-simple-stupid/89fccc82-840a-4a05-8b25-3e8dc951a1c8.html.

he can vividly recall the progressive ethos of an era: Dembicki, "Meet the Lawyer Trying to Make Big Oil Pay."

What Berman remembers most from high school: Lundegaard, "Keep It Simple, Stupid."

He knew some people in Highland Park who went from progressive to radical: Dembicki, "Meet the Lawyer Trying to Make Big Oil Pay."

"It's a very sad story," he later explained: Angeion Group, "Steve Berman—Rapid Fire Q&A," Leading Litigator Series, episode 13, September 7, 2016, www.youtube.com/watch?v=sK1jw4eDmiU.

Around this time Berman met with a parent whose child: Dembicki, "Meet the Lawyer Trying to Make Big Oil Pay."

The very first anti-smoking case went to trial in 1962: David Margolick, "'Tobacco' Its Middle Name, Law Firm Thrives, for Now," *New York Times*, November 20, 1992, www.nytimes.com/1992/11/20/news/tobacco-its-middle-name-law-firm-thrives-for-now.html.

This denial strategy reached an infamous crescendo: Philip J. Hilts, "Tobacco Chiefs Say Cigarettes Aren't Addictive," *New York Times*, April 15, 1994, www.nytimes.com/1994/04/15/us/tobacco-chiefs-say-cigarettes-aren-t-addictive.html.

chemist with R. J. Reynolds had acknowledged the "overwhelming" evidence: Scott Shane, "Tobacco Papers: Smoke, Secrets Companies Suppressed Their Covert Research, States' Suits Disclose," *Baltimore Sun*, September 28, 1997, www.baltimoresun.com/news/bs-xpm-1997-09-28-1997271017-story.html.

In the mid-1990s, a whistleblower went public: Marie Brenner, "The Man Who Knew Too Much," *Vanity Fair*, May 1996, www.vanityfair.com/magazine/1996/05/wigand199605.

One collection of eighty-one papers included: John Mintz and Saundra Torry, "Internal R.J. Reynolds Documents Detail Cigarette Marketing Aimed at Children," *Washington Post*, January 15, 1998, www.washingtonpost.com/wp-srv/national/longterm/tobacco/stories/memos1.htm.

Most of the previous legislation was brought on behalf of individuals: Dembicki, "Meet the Lawyer Trying to Make Big Oil Pay."

"The quiet, youthful [lawyer] worked brutal hours": Dan Zegart, *Civil Warriors: The Legal Siege on the Tobacco Industry* (London: Delta Publishing, 2001), 214.

while admitting that cigarettes cause cancer: David Usborne, "Smoking Kills: Tobacco Firm," *Independent*, March 21, 1997, www.independent.co.uk/news/smoking-kills-tobacco-firm-1274023.html.

Berman gave the opening statement at the Washington trial: Dembicki, "Meet the Lawyer Trying to Make Big Oil Pay."

Christine Gregoire said it marked the end of a deceitful era: "Tobacco a Done Deal," CNN Money, November 20, 1998, https://money.cnn.com/1998/11/20/companies/tobacco_deal/.

"Saudi Arabia of the western world"

They'd been alerted to a terrible smell: Michael Svoboda, "'Fatal Isolation' Delves Into Lethal 2003 Paris Heat Wave," *Yale Climate Connections*, August 27, 2015, https://yaleclimateconnections.org/2015/08/new-analysis-of-2003-fatal-paris-heat-wave/.

University of Oxford climate scientist Daniel Mitchell calculated: "Climate Change to Blame for Deaths in 2003 Heat Wave, New Study Says," *Inside Climate News*, July 8, 2016, https://insideclimatenews.org/news/08072016/climate-change-blame-deadliness-2003-heat-wave-new-study-paris-london/.

officially acknowledged that the oil sands could hold 180 billion barrels: Jeff Gerth, "Canada Builds a Large Oil Estimate on Sand," *New York Times*, June 18, 2003, www.nytimes.com/2003/06/18/business/canada-builds-a-large-oil-estimate-on-sand.html.

"No one disputes that the oil-sands industry has come of age": Cait Murphy, "The Big Dig Mining for Oil the Canadian Way," CNN Money, December 8, 2003, https://money.cnn.com/magazines/fortune/fortune_archive/2003/12/08/355120/index.htm.

producers opened up the Athabasca Oil Sands Project: "Explosion Halts

Production at Oilsands," Associated Press, January 6, 2003,https://apnews.com/article/fef3975fcc42be3e00431c23e8f265c2.

Suncor had just completed building Project Millennium: "Energy for the Future (1990s, 2000s, 2010s & 2020s)," Suncor, last updated 2022, www.suncor.com/en-ca/who-we-are/history/energy-for-the-future.

Imperial Oil officially opened a $650 million expansion: "Imperial's Mahkeses Facility Reflects Advances in Environmental Protection, Technology," *EcoWeek*, June 9, 2003, www.ecoweek.ca/issues/isarticle.asp?aid=1000175926.

The Wall Street Journal *described it as*: Tamsin Carlisle, "A Black-Gold Rush in Alberta," *Wall Street Journal*, September 15, 2005, www.post-gazette.com/business/businessnews/2005/09/15/A-black-gold-rush-in-Alberta/stories/200509150522.

Writing in the Nation, *Naomi Klein reported*: Naomi Klein, "Baghdad Burns, Calgary Booms," *Nation*, May 31, 2007, www.thenation.com/article/archive/baghdad-burns-calgary-booms/.

The boom was also happening south of the border: "Tar Sands: Feeding U.S. Refinery Expansion With Dirty Fuel," Environmental Integrity Project, June 2008, http://crgna.org/blog/wp-content/uploads/2013/08/Tar_Sand_Report.pdf.

bitumen still presented difficulties for the company owing to its high sulfur content: Richard Mial, "Refinery Takes Canadian Oil and Turns It Into Gasoline for Wisconsin," *La Crosse Tribune*, March 8, 2010, https://lacrossetribune.com/news/local/refinery-takes-canadian-oil-and-turns-it-into-gasoline-for-wisconsin/article_e76ad10e-2a6d-11df-8363-001cc4c002e0.html.

challenges of handling this gusher of heavy oil were very much worth it: Christopher Leonard, *Kochland: The Secret History of Koch Industries and Corporate Power in America* (New York: Simon & Schuster, 2019), 544.

Even that doesn't fully capture the scale: Anders Hayden, *When Green Growth Is Not Enough: Climate Change, Ecological Modernization, and Sufficiency* (Montreal: McGill-Queen's University Press, 2014).

a threat to the entire global climate system: Andrew Nikiforuk, "Dirty Oil: How the Tar Sands Are Fueling the Global Climate Crisis," Greenpeace, September 2009, https://climateactionnetwork.ca/wp-content/uploads/2014/02/dirtyoil.pdf.

"What Makes Weather?"

Bill McCaffrey, CEO *of the Canadian oil sands producer* MEG *Energy, explained*: "A Tale of Two Leaders: Sir John Browne and Bill McCaffrey," Petroleum History, www.petroleumhistory.ca/history/A%20Tale%20of%20Two%20Leaders%20-%20Sir%20John%20Browne%20and%20Bill%20McCaffrey.pdf.

containing potentially 3.7 billion barrels of oil: Mark Milner, "BP to Pump Billions Into Oil Sands Despite Green Worries and High Costs," *Guardian*, December 6, 2007, www.theguardian.com/business/2007/dec/06/bp.oil.

was expanding so that it could process up to 170,000 barrels: Chuck Marvin, "BP Alters Strategy With Oil Sands Deal," *The Street*, December 5, 2007, www.thestreet.com/investing/stocks/bp-alters-strategy-with-oil-sands-deal-10393088.

resulting in emissions anywhere from 14 to 20 percent higher: Lisa Song, "Exclusive Interview: Why Tar Sands Oil Is More Polluting and Why It Matters," *Inside Climate News*, May 22, 2012, www.reuters.com/article/idUS201043482520120522.

a twenty-four-minute documentary released in 1991: *This Earth—What Makes Weather?*, BP Video Library, January 1, 1991, www.bpvideolibrary.com/record/463.

The film was likely viewed by "hundreds if not thousands" of people: Vatan Hüzeir, "BP Knew the Truth About Climate Change 30 Years Ago. Now, It's Time to Ban Fossil Industry Advertising," *Follow the Money*, May 26, 2020, www.ftm.eu/articles/bp-video-climate-change-1990-engels.

directly contradicted the warnings contained in BP*'s documentary*: "Issues & Options: Potential Global Climate Change," Global Climate Coalition, 1994, www.climatefiles.com/denial-groups/global-climate-coalition-collection/1994-potential-global-climate-change-issues/.

"I was about to become an environmental activist": Hüzeir, "BP Knew the Truth About Climate Change 30 Years Ago."

During a May speech at Stanford that year: John Browne, "Climate Change Speech," BP, Stanford University, May 17, 1997, www.climatefiles.com/bp/bp-climate-change-speech-to-stanford/.

resulted in tons of positive publicity: Geoffrey Lean, "Stormy Ride for Two Unlikely Friends," *Independent*, August 23, 1997, www.independent.co.uk/news/stormy-ride-for-two-unlikely-friends-1247105.html.

It still remained a member of the American Legislative Exchange Council: "American Legislative Exchange Council (ALEC)," DeSmog, last updated 2021, www.desmog.com/american-legislative-exchange-council/.

Greenpeace responded to the deal by accusing BP: Terry Macalister, "Greenpeace Calls BP's Oil Sands Plan an Environmental Crime," *Guardian*, December 7, 2007, www.theguardian.com/business/2007/dec/07/bp.

"Global energy powerhouse"

Stéphane Dion looked uncomfortable: "Stéphane Dion CTV Interview Part 1," CTV, 2008, www.youtube.com/watch?v=K3X_BUUzkZ8.

the network's president, Robert Hurst, said in defense: Canadian Press, "CTV Defends Showing Dion Interview Fumbles," *Globe and Mail*, October 17, 2008, www.theglobeandmail.com/news/national/ctv-defends-showing-dion-interview-fumbles/article17972757/.

revenue raised from the tax would be used to lower people's income taxes: "Carbon Tax Plan 'Good for the Wallet,' Dion Pledges," CBC, June 19, 2008, www.cbc.ca/news/canada/carbon-tax-plan-good-for-the-wallet-dion-pledges-1.704607.

Harper was ruthless in his attacks on Dion's climate plan: Jennifer Graham, "Carbon Tax Would 'Screw' Canada: Harper," *Toronto Star*, June 21, 2008, www.thestar.com/news/canada/2008/06/21/carbon_tax_would_screw_canada_harper.html.

Harper's political views were formed by reading the libertarian canon: "Stephen Harper, a Cerebral Partisan," CBC News, March 16, 2006, www.cbc.ca/news2/background/harper_stephen/.

He had the cool demeanor of a strategist: Frances Russell, "Stephen Harper and the Triumph of the Corporation State," *iPolitics*, October 31, 2012, https://ipolitics.ca/2012/10/31/frances-russell-stephen-harper-and-the-triumph-of-the-corpora tion-state/.

J. Howard Pew had bonded with Alberta premier Ernest Manning: Darren Dochuk, "Prairie Fire: The New Evangelicalism and the Politics of Oil, Money, and Moral Geography," in *American Evangelicals and the 1960s*, ed. Axel R. Schäfer (Madison, WI: University of Wisconsin Press, 2013), 50.

a 2002 fundraising letter that Harper sent to Alliance supporters: "Harper's Letter Dismisses Kyoto as 'Socialist Scheme,'" CBC News, January 30, 2007, www.cbc.ca/news/canada/harper-s-letter-dismisses-kyoto-as-socialist-scheme-1.693166.

Harper penned an open letter in the Wall Street Journal: Stephen Harper, "Canadians Stand With You," *Wall Streel Journal*, March 28, 2003, www.wsj.com/articles/SB104881540524220000.

he said in 2006 when asked about climate change: News Staff, "Harper's Sincerity on Global Warming Questioned After Ex-Minister Assails Climate 'Alarmism,'" *City News*, February 25, 2010, https://toronto.citynews.ca/2010/02/25/harpers-sincerity-on-global-warming-questioned-after-ex-minister-assails-climate-alarmism/.

climate change deniers noted these comments approvingly: "An Open Letter to Prime Minister Stephen Harper," April 6, 2006, www.epw.senate.gov/public/index.cfm/press-releases-all?ID=1E639422-7094-4972-83AF-EE40EE302D41.

revealed a new diplomatic strategy for Canada: Mike De Souza, "Harper's Timeline: Canada on Climate Change from 2006-2014," *Narwhal*, September 19, 2014, https://thenarwhal.ca/harper-s-timeline-canada-climate-change-2006-2014/.

Harper referred to Canada: Jane Taber, "PM Brands Canada an 'Energy Superpower,'" *Globe and Mail*, July 15, 2006, www.theglobeandmail.com/news/national/pm-brands-canada-an-energy-superpower/article18167474/.

teamed up for a television ad: "Nancy Pelosi and Newt Gingrich Commercial on Climate Change," 2008, www.youtube.com/watch?v=qi6n_-wB154.

V: BLAME CANADA

"Back off dudes!"

the first-ever Black presidential contender was given a tour: Kate Sheppard, "Obama Heads to Nevada, Takes On McCain in Energy Policy Address," *Grist*, June 25, 2008, https://grist.org/article/nevada-say-never-again/.

his statements referring to oil were carefully parsed: Peter Foster, "Obama Plays 'Dirty' Oil Card," *National Post*, June 26, 2008, www.pressreader.com/canada/national-post-latest-edition/20080626/282604553605720.

They were joined in the meeting by executives: Tonda MacCharles, "Canada to Hard Sell Obama, McCain on Tar Sands," *Toronto Star*, August 21, 2008, www.pressreader.com/canada/toronto-star/20080827/28187011 4232735.

Enbridge contractors accidentally sawed down: Geoff Dembicki, "How Enbridge Sawed Off Good Relations With BC First Nations," *Tyee*, January 16, 2012, https://thetyee.ca/News/2012/01/16/Enbridge-First-Nations-Relations/.

U.S. policymakers were proposing climate legislation: Geoff Dembicki, "The Battle to Block Low Carbon Fuel Standards," *Tyee*, March 17, 2011, https://thetyee.ca/News/2011/03/17/LowCarbonFuelFight/.

While visiting a gas station in the Los Angeles neighborhood: Reuters staff, "Barack Obama Endorses Low Carbon Fuel Standard," Reuters, June 12, 2007, www.reuters.com/article/us-fuel-obama/barack-obama-endorses-low-carbon-fuel-standard-idUSN1229186820070612.

it would potentially strike a major blow to the oil sands: Dembicki, "The Battle to Block Low Carbon Fuel Standards."

extremely sensitive toward anything uttered about the oil sands: Domenick Yoney, "U.S. Mayors Call for Oil Sands Boycott," *Autoblog*, June 27, 2008, www.autoblog.com/2008/06/27/u-s-mayors-call-for-oil-sands-boycott/.

Harper's cabinet ministers began publicly "calling for a pact": Joseph Romm, "Harper Proposes Joint Climate Pact That Would Protect Alberta Oil Sands," *Grist*, November 19, 2008, https://grist.org/article/canada-tries-to-tar-sandbag-obama-on-climate/.

"A full-on barney"

Murdoch wanted the media company to start taking climate change seriously: Geoff Dembicki, "Rupert Murdoch Has Known We've Been in a Climate Emergency Since 2006, Documents Show," *VICE News*, September 23, 2021, www.vice.com/en/article/n7byqw/rupert-murdoch-climate-change-fox-news-news-corp.

"Bolt opened his comments by congratulating Gore": "Bolt's Minority View," ABC, October 30, 2006, www.abc.net.au/mediawatch/episodes/bolts-minority-view/9975752.

Bolt remembers the exchange differently: Dembicki, "Rupert Murdoch Has Known We've Been in a Climate Emergency Since 2006."

with the goal of becoming carbon-neutral by 2010: Rupert Murdoch, "Global Energy Initiative Launch," News Corp, May 9, 2007, https://web.archive.org/web/20130509133621/http://gei.newscorp.com/what/2007/05/global-energy-initiative-launc.html#more.

In the United States nearly 50 percent of respondents now agreed: Anthony Leiserowitz, "International Public Opinion, Perception, and Understanding of Global Climate Change," Yale University, July 2010, www.researchgate.net/publication/23866 7743_International_Public_Opinion_Perception_and_Understanding_of_Global_Climate_Change.

The CEO of Exxon, Rex Tillerson, seemed to acknowledge that new reality: Rex Tillerson, "The State of the Energy Industry: Strengths, Realities, and Solutions," CERAWeek 2007, February 14, 2007, www.climatefiles.com/exxonmobil/

2007-speech-exxonmobil-ceo-rex-tillerson-ceraweek/.

John McCain felt comfortable making it part of his presidential campaign: Elisabeth Bumiller and John M. Broder, "McCain Differs With Bush on Climate Change," *New York Times*, May 13, 2008, www.nytimes.com/2008/05/13/us/politics/12cnd-mccain.html.

convinced of the seriousness of climate change by his son James: Dembicki, "Rupert Murdoch Has Known We've Been in a Climate Emergency Since 2006."

During a rare 2007 interview with the environmental outlet Grist: Amanda Little, "An Interview With Rupert Murdoch About News Corp's New Climate Strategy," *Grist*, May 17, 2007, https://grist.org/article/murdoch1/.

Around this time, Fox Studios created a public service announcement: Dembicki, "Rupert Murdoch Has Known We've Been in a Climate Emergency Since 2006."

"Public embarrassment"

prime minister Stephen Harper called global warming: Stephen Harper, "Prime Minister Stephen Harper Calls for International Consensus on Climate Change," Government of Canada, June 4, 2007, www.canada.ca/en/news/archive/2007/06/prime-minister-stephen-harper-calls-international-consensus-climate-change.html.

Harper had sent out a fundraising letter deeming the Kyoto Protocol: "Harper's Letter Dismisses Kyoto as 'Socialist Scheme,'" CBC News, January 30, 2007, www.cbc.ca/news/canada/harper-s-letter-dismisses-kyoto-as-socialist-scheme-1.693166.

claimed Tom Harris, who had helped organize the Imperial Oil event: Peter Gorrie, "Who's Still Cool on Global Warming?," *Toronto Star*, January 28, 2007, www.thestar.com/news/2007/01/28/whos_still_cool_on_global_warming.html.

the Conservative leader did everything he could to suppress the science: Andrew Cuddy, "Troubling Evidence: The Harper Government's Approach to Climate Science Research in Canada," Climate Action Network Canada, March 2010, https://web.archive.org/web/20211104030307/https://www.bcsea.org/sites/default/files/Troubling-Evidence-The-Harper-Governments-Approach-to-Climate-Science_Research_in_Canada.pdf.

Harper was appointing climate change deniers to key scientific positions: Mitchell Anderson, "The Stephen Harper War on Climate Science," DeSmog, May 11, 2019, www.desmog.com/2009/05/11/stephen-harper-war-climate-science/.

"They're struggling forward"

was a bad day for an American named Michael Whatley: All the references in this chapter to Michael Whatley and Gary Mar's campaign against low-carbon fuel standards are taken from: Geoff Dembicki, "Big Oil and Canada Thwarted U.S. Carbon Standards," *Salon*, December 15, 2011, www.salon.com/2011/12/15/big_oil_and_canada_thwarted_u_s_carbon_standards/.

the policy that had been first adopted by California in 2007: Debra Khan, "California Adopts Low-Carbon Fuel Standard," *Scientific American*, April 24, 2009, www.scientificamerican.com/article/california-adopts-low-car/#.

That attempt ran into fierce opposition: Geoff Dembicki, "Canada Teams With Oil Lobby to Fight US Clean Energy Clause," *Tyee*, March 16, 2011, https://thetyee.ca/News/2011/03/16/Canada_Teams_With_Oil_Lobby/.

"Global warming!"

It was early 2010, and Barack Obama was attempting to rally Americans: Eric Pooley, *The Climate War: True Believers, Power Brokers, and the Fight to Save the Earth* (New York: Hachette Books, 2010).

News Corp reached out to Luntz: Pooley, *The Climate War*, 98.

The Pine Bend Refinery in Minnesota refined roughly 25 percent: Geoff Dembicki, "US Tea Party's Deep Ties to Oil Sands Giant," *Tyee*, November 1, 2010, https://thetyee.ca/News/2010/11/01/TeaPartyTies/.

Koch Industries contributed heavily to groups that spread disinformation: "Koch Industries Secretly Funding the Climate Denial Machine," Greenpeace, March 2010, www.greenpeace.org/usa/wp-content/uploads/legacy/Global/usa/

report/2010/3/koch-industries-secretly-fund.pdf?9e7084.

a TV advertisement featuring a rich young man next to a plate of canapés: Brad Johnson, "Front Group for Polluter Billionaires Wastes $140K on Goofy Global Warming Denial Ads," *Think Progress*, February 11, 2009, https://archive.thinkprogress.org/front-group-for-polluter-billionaires-wastes-140k-on-goofy-global-warming-denial-ads-a4af63f2738f/.

a speech he gave in October 2009 at the Crystal Gateway Marriott: Jane Mayer, *Dark Money: The Hidden History of the Billionaires Behind the Rise of the Radical Right* (New York: Anchor Books, 2017), 241.

An unnamed Republican insider quoted by Jane Mayer: Jane Mayer, "Covert Operations," *New Yorker*, August 23, 2010, www.newyorker.com/magazine/2010/08/30/covert-operations.

The victors included Minnesota representative: Kate Galbraith, "Michele Bachmann Seeks 'Armed and Dangerous' Opposition to Cap-and-Trade," *New York Times*, March 25, 2009, https://green.blogs.nytimes.com/2009/03/25/michele-bachmann-seeks-armed-and-dangerous-opposition-to-cap-and-trade/.

started urging protesters to turn their anger against the cap-and-trade: Mayer, *Dark Money*, 266.

Ryan Hecker proposed that activists vote online: Teddy Davis, "Tea Party Activists Unveil 'Contract From America,'" ABC News, April 14, 2010, https://abcnews.go.com/Politics/tea-party-activists-unveil-contract-america/story?id=10376437.

Graham began to waver in his support: Lee Fang, "Graham Reportedly Terrified That Fox News Would Learn That He Was Negotiating a Climate Bill," *Think Progress*, October 4, 2010, https://archive.thinkprogress.org/graham-reportedly-terrified-that-fox-news-would-learn-that-he-was-negotiating-a-climate-bill-4797ea1feb52/.

Rupert Murdoch had stated his belief that he could convince Fox News hosts: Geoff Dembicki, "Rupert Murdoch Has Known We've Been in a Climate Emergency Since 2006, Documents Show," *VICE News*, September 23, 2021, www.vice.com/en/article/n7byqw/rupert-murdoch-climate-change-fox-news-news-corp.

Sammon wrote in an email to Fox News staffers: Keach Hagey, "Fox Editor Urged Climate Skepticism," *Politico*, December 15, 2010, www.politico.com/story/2010/12/fox-editor-urged-climate-skepticism-046409.

An article on the Fox News site helpfully informed viewers: Associated Press, "John McCain Links Climate Policy to Terror Threat in Energy Speech," Fox News, April 23, 2007, www.foxnews.com/story/john-mccain-links-climate-policy-to-terror-threat-in-energy-speech.amp.

by 2010, Fox News was referring to: Deneen Borelli, "Another Tea Party Victory," Fox News, May 3, 2010, www.foxnews.com/opinion/another-tea-party-victory.

Murdoch's public views on climate change also appeared to shift: Dembicki, "Rupert Murdoch Has Known We've Been in a Climate Emergency Since 2006."

VI: THE CLIMATE GOES TO COURT

"The island is sad that it's going away"

wanted to sue the oil and gas industry: Geoff Dembicki, "Meet the Lawyer Trying to Make Big Oil Pay for Climate Change," VICE, December 22, 2017, www.vice.com/en/article/43qw3j/meet-the-lawyer-trying-to-make-big-oil-pay-for-climate-change.

It estimated that the cost of relocation could be up to $400 million: Max Ajl, "Alaska's Soon-to-Be Climate Refugees Sue Energy Companies for Relocation," *Inside Climate News*, June 16, 2009, https://insideclimatenews.org/news/16062009/alaskas-soon-be-climate-refugees-sue-energy-companies-relocation/.

the complaint accused those companies of deliberately spreading doubt: Attorneys for Plaintiffs, "Kivalina Complaint," February 26, 2008, http://blogs2.law.columbia.edu/climate-change-litigation/wp-content/uploads/sites/16/case-documents/2008/20080226_docket-408-cv-01138-SBA_complaint.pdf.

the fossil fuel industry had intentionally deceived the public: Stephen Faris, "Conspiracy Theory," *Atlantic*, June 2008, www.theatlantic.com/magazine/archive/2008/06/conspiracy-theory/306812/.

Despair was pervasive: Dembicki, "Meet the Lawyer Trying to Make Big Oil Pay for Climate Change."

Lucy Adams told one visitor that the town was becoming unrecognizable: Marissa Knodel, "Conceptualizing Climate Justice in Kivalina," *Seattle University Law Review* 37 (2014), https://digitalcommons.law.seattleu.edu/cgi/viewcontent.cgi?article=2243&context=sulr.

Berman had history with Susman: Dembicki, "Meet the Lawyer Trying to Make Big Oil Pay for Climate Change."

Their complaint accused prominent oil and gas companies: Attorneys for Plaintiffs, "Kivalina Complaint."

It was a novel legal argument: Felicity Barringer, "Alaskan Village Files Suit Against Energy Companies," *New York Times*, February 27, 2008, www.nytimes.com/2008/02/27/world/americas/27iht-alaska.1.10468317.html.

a U.S. district court ruled against the Kivalina lawsuit: United States District Court for the Northern District of California, "Order Granting Defendants' Motions to Dismiss," September 30, 2009, www.informea.org/sites/default/files/court-decisions/COU-158012.pdf.

"A way to justify exploitation"

The complaint listed contrarians such as Patrick Michaels: Attorneys for Plaintiffs, "Kivalina Complaint," February 26, 2008, http://blogs2.law.columbia.edu/climate-change-litigation/wp-content/uploads/sites/16/case-documents/2008/20080226_docket-408-cv-01138-SBA_complaint.pdf.

a short biography on the Institute of Economic Affairs website notes: John Blundell, "The Life and Work of Sir Antony Fisher," Institute of Economic Affairs, July 10, 2013, https://iea.org.uk/blog/the-life-and-work-of-sir-antony-fisher.

Fisher dedicated special attention in the early days: Gerald Frost, "Antony Fisher: Champion of Liberty," Institute of Economic Affairs, 2008, https://mannkal.org/downloads/guests/antonyfisher.pdf.

the institute published a 180-page book: Laura Jones, *The Science and Politics of Global Warming* (Vancouver: Fraser Institute, 2007), www.fraserinstitute.org/sites/default/files/ScienceandPoliticsGlobalWarming.pdf.

according to investigative journalist Donald Gutstein: Donald Gutstein, "Following the Money: The Fraser Institute's Tobacco Papers," October 14, 2009, http://donaldgutstein.com/press/following-the-money-the-fraser-institutes-tobacco-papers/.

was short on cash and actively seeking funding from cigarette makers: "Fraser Institute," Tobacco Tactics, last updated February 5, 2020, https://tobaccotactics.org/wiki/fraser-institute/.

better luck getting checks from oil and gas companies: "Fraser Institute," DeSmog, last updated 2021, www.desmog.com/fraser-institute/.

noted Carleton University researcher Aldous Sperl in his dissertation: Aldous Sperl, "Climate Change Denial in Canada: An Evaluation of the Fraser Institute and Friends of Science Positions," Master of Arts in Geography thesis for Carleton University, 2013, https://curve.carleton.ca/system/files/etd/4a4607a0-77e4-44a3-8494-a3114c479e74/etd_pdf/5a11fe302187d37f3acf9774da0dae7d/sperl-climatechangedenialincanadaanevaluationof.pdf.

Actual climate experts slammed the report: Richard Littlemore, "DeSmog Leaks Advance Copy of Think Tank's IPCC Attack," DeSmog, January 31, 2007, www.desmog.com/2007/01/31/desmog-leaks-advance-copy-of-think-tanks-ipcc-attack/.

The institute saw coverage like this as one of its major reasons for existing: "35 Big Ideas: How the Fraser Institute Is Changing the World," Fraser Institute, 2009, www.fraserinstitute.org/sites/default/files/35th-anniversary-book.pdf.

Soon the Fraser Institute was getting big checks from Koch Industries: Alexis Stoymenoff, "'Charitable' Fraser Institute Accepted $500k in Foreign Funding From Koch Oil Billionaires," *Vancouver Observer*, April 26, 2012, https://web.archive.org/web/20150708060846/http://www.vancouverobserver.com/

politics/2012/04/25/charitable-fraser-institute-accepted-500k-foreign-funding-oil-billionaires.

contributed nearly $1.7 million to the Fraser Institute: "Fraser Institute," DeSmog.

That is not how the think tank presents itself to the public: "Our Mission," Fraser Institute, last updated 2022, www.fraserinstitute.org/about.

counters Connor Gibson, a former researcher with Greenpeace USA: Geoff Dembicki, "The Emperor Kenney's New Clothes," *Tyee*, April 24, 2020, https://thetyee.ca/Analysis/2020/04/24/The-Emperor-Kenneys-New-Clothes/.

"I remember being angry every day"

But Joanna Sustento felt like the world was crashing down on her: Joanna Sustento, "The Impossible—A Story About Love, Decisions and Survival. A Story About Haiyan," Storya.ph, July 30, 2014, https://web.archive.org/web/20190812195318/http://www.storya.ph/joanna-sustento-story-53ce07e6b9769.

they tried to get Julius medical treatment for his injured leg: Joanna Sustento, interview with author, November 15, 2021.

Sustento quickly rejected that explanation: Geoff Dembicki, "The Woman Going After Big Energy for the Typhoon That Killed Her Family," *VICE*, February 14, 2018, www.vice.com/en/article/3k7dv9/the-woman-going-after-big-energy-for-the-typhoon-that-killed-her-family.

Haiyan registered 8.1 on the Dvorak intensity scale: Eric Holthaus, "Super Typhoon Haiyan, One of the Strongest Storms Ever Seen, Hit the Philippines With Record Force," *Quartz*, November 7, 2013, https://qz.com/144734/super-typhoon-haiyan-one-of-the-strongest-storms-ever-seen-is-about-to-hit-the-philippines/.

BP in its 1991 educational video on climate change had warned: *This Earth—What Makes Weather?*, BP Video Library, January 1, 1991, www.bpvideolibrary.com/record/463.

VII: WELL-OILED ALLIES

"Stacked with friends"

Tillerson entered through the back door: Shane Goldmacher, Josh Dawsey, and Matthew Nussbaum, "Why Trump Picked Rex Tillerson," *Politico*, December 13, 2016, www.politico.com/story/2016/12/rex-tillerson-donald-trump-secretary-of-state-232581.

In Canada's oil sands, people were closely watching: John Ibbitson, Jeffrey Jones, and Jeff Lewis, "Rex Tillerson, Trump's Likely Secretary of State, Is a Life-Long Oil Man Who Backs Keystone," *Globe and Mail*, December 11, 2016, www.theglobeandmail.com/news/world/trumps-likely-secretary-of-state-a-life-long-oil-man-who-backs-keystone/article33295132/.

he shared a table with Canadian ambassador Gary Doer: Alexander Panetta, "Two Texans Who Love Canadian Oil Offered Pipeline-Policy Roles in Trump Cabinet," *Calgary Herald*, December 13, 2016, https://calgaryherald.com/business/energy/two-texans-who-love-canadian-oil-offered-pipeline-policy-roles-in-trump-cabinet.

and thereby raise the value of Exxon's investments: "Policy Brief," Greenpeace, March 2017, www.greenpeace.org/usa/wp-content/uploads/2017/03/PolicyBrief-TillersonKXLRecusal.pdf.

Conway had accepted an invitation from the Alberta Prosperity Fund: Alberta Prosperity Fund, "Conway Visit Cancelled," Cision, January 7, 2017, www.newswire.ca/news-releases/conway-visit-cancelled-609980445.html.

The good news for the oil sands just kept on coming: Panetta, "Two Texans Who Love Canadian Oil Offered Pipeline-Policy Roles in Trump Cabinet."

Tillerson apparently wasn't all that concerned about the climate impacts: Associated Press, "Climate Change Fears Overblown, Says ExxonMobil Boss," *Guardian*, June 28, 2012, www.theguardian.com/environment/2012/jun/28/exxonmobil-climate-change-rex-tillerson.

refunds had to be given out when she canceled at the last minute: Alberta Prosperity Fund, "Conway Visit Cancelled."

"Friends in unexpected places"

Canadian prime minister Justin Trudeau was there to welcome them: Julian Robinson, "Canadian PM Justin Trudeau Shows the World—And Donald Trump—'How to Open Your Heart' as He Greets Syrian Refugees Arriving in Toronto," *Daily Mail*, December 11, 2015, www.dailymail.co.uk/news/article-3355482/Canadas-prime-minister-welcome-refugees-airport.html.

the Trudeau government actually saw Trump becoming president as a good thing: Martin Lukacs, "Revealed: Trudeau Government Welcomed Oil Lobby Help for US Pipeline Push," *Guardian*, February 9, 2018, www.theguardian.com/environment/true-north/2018/feb/09/trudeau-government-welcomed-oil-lobby-help-for-us-pipeline-push-documents.

a "deeply disappointed" Trudeau released a statement: Peter Zimonjic, "Trudeau Tells Trump Canada Is Disappointed by Withdrawal From Paris Climate Deal," CBC News, June 1, 2017, www.cbc.ca/news/politics/trudeau-mckenna-trump-paris-deal-1.4142211.

the joy of getting Harper out was intoxicating: Geoff Dembicki, "Will Joe Biden Betray the Climate Movement Like Justin Trudeau Did?," *New Republic*, December 31, 2020, https://newrepublic.com/article/160709/joe-biden-betray-climate-movement-like-justin-trudeau.

his predecessor Michael Ignatieff saw a "political animal": Sonja Puzic, "Trudeau 'Has the Skills' to Take On Harper, Ignatieff Says," CTV News, September 29, 2013, www.ctvnews.ca/politics/trudeau-has-the-skills-to-take-on-harper-ignatieff-says-1.1475560.

This apparently surprised some of the people gathered to hear him speak: Megan Fitzpatrick, "Justin Trudeau Shares 'Steadfast' Keystone XL Support in D.C.," CBC News, October 25, 2013, www.cbc.ca/news/world/justin-trudeau-shares-steadfast-keystone-xl-support-in-d-c-1.2251745.

Trudeau flew to Calgary and gave a speech to a room of oil executives: Justin Trudeau, "Liberal Party of Canada Leader Justin Trudeau's Speech to the Calgary Petroleum Club," Liberal Party of Canada, October 30, 2013, https://liberal.ca/liberal-party-canada-leader-justin-trudeaus-speech-calgary-petroleum-club/.

Summing up the mood, the media outlet Grist *wrote*: Daniel Kessler, "Harper in Denial: Stephen Harper's Refusal of Climate Reality," *Grist*, December 13, 2013, https://grist.org/article/harper-in-denial-stephen-harpers-refusal-of-climate-reality/.

Trudeau at one point tweeted a photo of himself in a canoe: Geoff Dembicki, "How Trudeau's Broken Promises Fuel the Growth of Canada's Right," *Tyee*, September 4, 2019, https://thetyee.ca/Analysis/2019/09/04/Trudeau-Broken-Promises-Fuel-Right/.

One of Trudeau's top policy advisers leading up to the election was Cyrus Reporter: Daniel Leblanc, "Former Top Trudeau Aide Joins Law Firm to Help Clients 'Navigate' Government Regulations," *Globe and Mail*, February 13, 2017, www.theglobeandmail.com/news/politics/top-pmo-staffer-joins-law-firm-to-help-clients-navigate-government-regulations/article34009184/.

Trudeau's campaign cochair was Dan Gagnier: Laura Payton, "Liberals Knew About TransCanada Work, Says Gagnier," *Maclean's*, October 15, 2015, www.macleans.ca/politics/ottawa/liberals-knew-about-transcanada-work-former-campaign-co-chair/.

Yet climate groups across the country celebrated as the election results came in: Dembicki, "Will Joe Biden Betray the Climate Movement Like Justin Trudeau Did?"

Trudeau approved a $1.26 billion liquid natural gas plant: Michael Harris, "Trudeau Stuns Environmentalists With Dubious LNG Plant Approval," *iPolitics*, March 20, 2016, https://ipolitics.ca/2016/03/20/trudeau-stuns-environmentalists-with-dubious-lng-approval/.

Members of the Tsleil-Waututh First Nation didn't see it that way: John Paul Tasker, "Trudeau Cabinet to Discuss Trans Mountain Pipeline Tuesday as B.C. First Nation Vows to Oppose It," CBC News, November 28, 2016, www.cbc.ca/news/politics/trans-mountain-bc-pipeline-firstnations-1.3870838.

Berman came to believe that the country's environmental movement: Dembicki, "Will Joe Biden Betray the Climate Movement Like Justin Trudeau Did?"

were planning to nearly double bitumen production in northern Alberta: David Hughes, "Can Canada Expand Oil and Gas Production, Build Pipelines and Keep Its Climate Change Commitments?" Parkland Institute, June 2, 2016, www.parklandinstitute.ca/can_canada_expand.

"It just kept going and going"

Scott Pruitt was apparently a terrible tenant: Eliana Johnson, "Lobbyist Couple Had to Change the Locks on Pruitt," *Politico*, April 6, 2018, www.politico.com/story/2018/04/06/pruitt-was-the-kato-kaelin-of-capitol-hill-505658.

he approved a major oil sands pipeline expansion proposed by Enbridge: Luis Sanchez, "EPA Reportedly Approved a Company's Project While Pruitt Stayed in One of Its Lobbyists' Condos," *The Hill*, April 2, 2018, https://thehill.com/homenews/administration/381358-epa-approved-a-companys-project-while-the-epa-chief-rented-a-condo.

the Sierra Club claimed in a report on the project: "All Risk, No Reward: The Alberta Clipper Tar Sands Pipeline Expansion," Sierra Club, April 2014, https://content.sierraclub.org/creative-archive/sites/content.sierraclub.org.creative-archive/files/pdfs/0679-AlbertaClipperReport_09_web_0.pdf.

green groups put up satirical posters of Pruitt: Timothy Cama, "Environmental Group Hangs Posters Mocking Pruitt's DC Condo Rental," *The Hill*, April 6, 2018, https://thehill.com/policy/energy-environment/381948-environmental-group-hang-posters-mocking-pruitts-rental-in-dc.

the Tea Party Patriots urged its over 190,000 followers: Coral Davenport and Lisa Friedman, "Growing Crisis Threatens Scott Pruitt, E.P.A. Chief, as Top Aides Eye the Exits," *New York Times*, April 5, 2018, www.nytimes.com/2018/04/05/climate/epa-chief-scott-pruitt-pressure.html.

Charles and David Koch had publicly refused to back Trump's campaign: Kenneth P. Vogel and Cate Martel, "The Kochs Freeze Out Trump," *Politico*, July 29, 2015, www.politico.com/story/2015/07/kochs-freeze-out-trump-120752.

was key to Trump eventually becoming president: Lee Fang, "David Koch's Most Significant Legacy Is the Election of Donald Trump," *Intercept*, August 26, 2019, https://theintercept.com/2019/08/26/david-koch-donald-trump/.

Trump soon returned the favor: Jane Mayer, "The Danger of President Pence," *New Yorker*, October 16, 2017, www.newyorker.com/magazine/2017/10/23/the-danger-of-president-pence.

Pruitt coauthored an opinion piece in the National Review: Coral Davenport and Eric Lipton, "Trump Picks Scott Pruitt, Climate Change Denialist, to Lead E.P.A.," *New York Times*, December 7, 2016, www.nytimes.com/2016/12/07/us/politics/scott-pruitt-epa-trump.html.

A spokesperson for Koch Industries suggested to the New York Times: Hiroko Tabuchi, "The Oil Industry's Covert Campaign to Rewrite American Car Emissions Rules," *New York Times*, December 13, 2018, www.nytimes.com/2018/12/13/climate/cafe-emissions-rollback-oil-industry.html.

the billionaires were thrilled: "Efforts in Government: Advancing Principled Public Policy," Koch Seminar Network, 2018, www.documentcloud.org/documents/4364737-Koch-Seminar-Network.html.

the EPA gave to Koch Industries its "partner of the year" award: Press Release, "Flint Hills Resources Pine Bend Refinery, a Subsidiary of Koch Industries, Named 2017 Energy Star Partner of the Year," Flint Hills Resources, April 26, 2017, https://pinebendrefinery.com/wp-content/uploads/2017/09/4-26-17-Pine-Bend-Energy-Star-POY-release.pdf.

"This is an avalanche"

announced they were going to sue Big Oil: "San Francisco and Oakland Sue Top Five Oil and Gas Companies Over Costs of Climate Change," City Attorney of San Francisco, September 19, 2017, www.sfcityattorney.org/2017/09/19/san-francisco-oakland-sue-top-five-oil-gas-companies-costs-climate-change/.

Berman saw a lawsuit against Big Oil as one way to fight back: Geoff Dembicki, "Meet the Lawyer Trying to Make Big Oil Pay for Climate Change," VICE, December 22, 2017, www.vice.com/en/article/43qw3j/meet-the-lawyer-trying-to-make-big-oil-pay-for-climate-change.

"It was raining fire from the sky": Madison Park, "California Wildfires Evacuee: 'It Was Raining Fire From the Sky,'" CNN, July 26, 2016, www.cnn.com/2016/07/26/us/california-wildfires/index.html.

an attorney who defends oil and gas companies told the Washington Post: Dembicki, "Meet the Lawyer Trying to Make Big Oil Pay for Climate Change."

The legal complaint his firm served: "Complaint for Public Nuisance," City Attorney of San Francisco, September 19, 2017, www.sfcityattorney.org/wp-content/uploads/2017/09/2017-09-19-File-Stamped-Complaint-for-Public-Nuisance.pdf.

might not be such a stretch: Dembicki, "Meet the Lawyer Trying to Make Big Oil Pay for Climate Change."

A report released in November 2017: Center for International Environmental Law, "Smoke and Fumes: The Legal and Evidentiary Basis for Holding Big Oil Accountable for the Climate Crisis," November 2017, www.ciel.org/wp-content/uploads/2019/01/Smoke-Fumes.pdf.

He referenced them directly in the complaint: "Complaint for Public Nuisance," City Attorney of San Francisco.

"We are the beating heart"

It had helped shape the thinking of Republican presidents: George W. Bush, "Remarks to the Manhattan Institute in New York City," American Presidency Project, November 13, 2008, www.presidency.ucsb.edu/documents/remarks-the-manhattan-institute-new-york-city.

That September night Kenney received a gushing introduction: Geoff Dembicki, "The Emperor Kenney's New Clothes," *Tyee*, April 24, 2020, https://thetyee.ca/Analysis/2020/04/24/The-Emperor-Kenneys-New-Clothes/.

Up on stage, Kenney set the tone: Markham Hislop, "Jason Kenney Energy Speech Mimics Donald Trump, Vows to Retaliate Against Enemies of Alberta Oil and Gas if He Becomes Premier," *Energi Media*, October 26, 2018, https://energi.media/markham-on-energy/jason-kenney-energy-speech-mimics-donald-trump-vows-to-retaliate-against-enemies-of-alberta-oil-and-gas-if-he-becomes-premier/.

it had been thoroughly debunked numerous times: Markham Hislop, "Debunked: Vivian Krause's Tar Sands Campaign Conspiracy Narrative," *Energi Media*, May 14, 2019, https://energi.media/deep-dives/debunked-vivian-krauses-tar-sands-campaign-conspiracy-narrative/.

the oil sands were doing relatively well: Ian Hussey, "The Future of Alberta's Oil Sands Industry," Parkland Institute, March 10, 2020, www.parklandinstitute.ca/the_future_of_albertas_oil_sands_industry.

But by other measures the industry was in serious trouble: Smith Brain Trust, "Why Shell Has All But Given Up On Canada's Oil Sands," Robert H. Smith School of Business at the University of Maryland, March 15, 2017, www.rhsmith.umd.edu/news/why-shell-has-all-given-canadas-oil-sands#.

A week before Kenney's speech to the Relaunch conference: Hislop, "Jason Kenney Energy Speech Mimics Donald Trump."

The heads of these companies were becoming bitter and conspiratorial: Geoff Dembicki, "The Calgary Oil and Gas Execs Backing Scheer Might Surprise You," *Tyee*, August 14, 2019, https://thetyee.ca/Analysis/2019/08/14/Calgary-Oil-Gas-Execs-Backing-Scheer/.

Kenney was not fond—at least in public—of these comparisons: Geoff Dembicki, "A Political Grid Humming With Hatred: Who Taps Its Energy?" *Tyee*, April 29, 2019, https://thetyee.ca/Analysis/2019/04/29/Political-Grid-Humming-Hatred/.

Kenney boasted during his acceptance speech: "Read Jason Kenney's Prepared Victory Speech in Full After UCP Wins Majority in Alberta Election," *National Post*, April 17, 2019, https://nationalpost.com/news/canada/read-jason-kenneys-prepared-victory-speech-in-full-after-ucp-wins-majority-in-alberta-election.

was virtually identical to its own corporate logo: Bill Kaufmann, "Second Logo for Alberta's Energy War Room Comes Under Fire From U.S. Tech Firm," *Calgary Herald*, December 27, 2019, https://calgaryherald.com/news/local-news/alberta-governments-war-room-possibly-runs-afoul-over-second-logo.

the Canadian Energy Centre was forced to backtrack: Canadian Press, "Alberta Energy 'War Room' Chief Apologizes for Tweets Attacking *New York Times*," CBC News, February 12, 2020, www.cbc.ca/news/canada/calgary/alberta-energy-war-room-tom-olsen-jason-kenney-energy-1.5461950.

but his "free-market" strategy was also a flop: Dembicki, "The Emperor Kenney's New Clothes."

"They surrounded me"

Joanna Sustento showed up to Shell's headquarters in Bonifacio Global City: Lia Savillo, "We Met the Filipino Millennial Who Staged a Lone Protest for Climate Justice," *VICE News*, September 20, 2019, www.vice.com/en/article/zmjzqx/joanna-sustento-climate-change-activist-haiyan-typhoon-survivor-shell-protest-philippines.

she had a written message to Shell's CEO: Joanna Sustento, "A Letter to Shell for Taking Away My Family and Devastating My Community," Greenpeace, September 22, 2019, www.greenpeace.org/international/story/24426/a-letter-to-shell-for-taking-away-my-family-and-devastating-my-community/.

she never could have imagined being hauled away by police: Joanna Sustento, interview with author, November 15, 2021.

Sustento accompanied the actress Lucy Lawless to the Barents Sea: Press Release, "Lucy Lawless Joins Climate Change Survivor in Protest Against Statoil's Arctic Oil Exploitation," Greenpeace, July 22, 2017, www.greenpeace.org/aotearoa/press-release/lucy-lawless-joins-climate-change-survivor-in-protest-against-statoils-arctic-oil-exploitation/.

pursuing an unprecedented legal challenge to the oil and gas companies: Geoff Dembicki, "The Woman Going After Big Energy for the Typhoon That Killed Her Family," *VICE*, February 14, 2018, www.vice.com/en/article/3k7dv9/the-woman-going-after-big-energy-for-the-typhoon-that-killed-her-family.

"manufactured, produced, and sold fossil fuel products for decades": Geoff Dembicki, "It's Clear Who's to Blame for Australia's Fires, and It's Not Arsonists," *VICE*, January 13, 2020, www.vice.com/en/article/akwgp8/why-big-oil-should-pay-for-australian-fires.

The petition submitted by Greenpeace: Dembicki, "The Woman Going After Big Energy for the Typhoon That Killed Her Family."

Commission on Human Rights held a hearing in New York: Ucilia Wang, "Human Rights Hearing: Emotional Testimony, No Oil Industry Response," *Climate Docket*, September 28, 2018, www.climatedocket.com/2018/09/28/human-rights-hearing-philippines-new-york/.

The commission presented its final ruling: Guest, "Carbon Majors Can Be Held Liable for Human Rights Violations, Philippines Commission Rules," DeSmog, December 9, 2019, www.desmog.com/2019/12/09/carbon-majors-climate-liable-human-rights-violations-philippines-commission/.

This was around the time she got detained by police in Manila: Savillo, "We Met the Filipino Millennial Who Staged a Lone Protest for Climate Justice."

VIII: THE RIGHT TO LIVE

"Robbed of their options"

she started to develop vertigo and get bad migraines: Geoff Dembicki, "This 'Woke' Oil Company Has Been Illegally Polluting a Poor Latino Community," *VICE News*, May 17, 2021, www.vice.com/en/article/n7bndx/suncor-oil-company-illegally-polluting-denver-poor-latino-community.

A Harvard study from April 2020: "Air Pollution Linked With Higher COVID-19 Death Rates," Harvard T. Chan School of Public Health, May 5, 2020, www.hsph.harvard.edu/news/hsph-in-the-news/air-pollution-linked-with-higher-covid-19-death-rates/.

Molina became one of the refinery's most outspoken critics: Catie Cheshire, "These Commerce City Activists Lost the Election, but

Their Fight Is Just Beginning," *Westword*, November 9, 2021, www.westword.com/news/colorado-environment-denver-pollution-commerce-city-election-globeville-12614632.

It took direct aim at the Commerce City Refinery: "Complaint and Jury Demand," Boulder County, April 17, 2018, https://assets.bouldercounty.org/wp-content/uploads/2018/04/climate-accountability-lawsuit-filed-boulder-district-court.pdf.

thought Big Oil was more concerned than it was letting on: Geoff Dembicki, "Has Suncor Seen the Climate Crisis Coming for 61 Years?," *Tyee*, July 21, 2020, https://thetyee.ca/News/2020/07/21/Did-Suncor-See-Climate-Crisis-Coming/.

demanding a five-hour tutorial on the history and impacts: Umair Irfan, "A Federal Judge in a Climate Change Lawsuit Is Forcing Oil Companies to Cough Up Internal Documents," *Vox*, May 29, 2018, www.vox.com/2018/5/25/17394468/climate-lawsuits-san-francisco-oakland-alsup-exxon-bp.

But that wasn't the end of it: Geoff Dembicki, "San Francisco Leads Legal Effort Pressing Big Oil to Pay for Climate Impacts," *Audubon*, Fall 2021, www.audubon.org/magazine/fall-2021/san-francisco-leads-legal-effort-pressing-big-oil.

There is evidence to suggest Sun Oil received this warning: Dembicki, "Has Suncor Seen the Climate Crisis Coming for 61 Years?"

passed off the heavy costs of disasters and extreme weather: Chris McGreal, "Climate Crisis Has Cost Colorado Billions—Now It Wants Oil Firms to Pick Up the Bill," *Guardian*, August 2, 2021, www.theguardian.com/environment/2021/aug/02/climate-crisis-boulder-colorado-lawsuit-exxonmobil-suncor.

Molina was seeing the impacts of that first-hand: "Force of Nature Spotlight—Lucy Molina," 350 Colorado, December 28, 2020, www.facebook.com/350Colorado/videos/725917788362768/.

other municipalities and states across the U.S. added their names: "Climate Liability Litigation," Pay Up Climate Polluters, last updated 2021, https://payupclimatepolluters.org/cases.

"Why wouldn't I choose the right thing to do?"

Rosero was not one to doubt the motives of his employer: Enrique Rosero, interview by author, October 2, 2020. All quotes from Rosero in this chapter are taken from this interview.

Exxon's then CEO Lee Raymond had aggressively denied: Lee Raymond, "Energy—Key to Growth and Better Environment for Asia-Pacific Nations," World Petroleum Congress, October 13, 1997, www.climatefiles.com/exxonmobil/1997-exxon-lee-raymond-speech-at-world-petroleum-congress/.

In 2008, it gave $76,106 for a project by astrophysicist Willie Soon: Lizzie Plaugic, "ExxonMobil Funded a Climate Change Denier Years After It Claimed to Have Stopped," *The Verge*, February 27, 2015, www.theverge.com/2015/2/27/8122913/exxonmobil-climate-change-denier-willie-soon.

In 2009 alone, Exxon contributed $1.3 million: Kert Davies, "Exxon Continued to Fund Climate Denial in 2009," Greenpeace, July 19, 2010, www.greenpeace.org/usa/exxon-continued-to-fund-climate-denial-in-2009/.

Rosero's work helped make possible a development plan: Geoff Dembicki, "How Exxon Silences Staff Alarmed by the Climate Crisis, According to a Former Employee," *VICE News*, October 29, 2020, www.vice.com/en/article/bvx84q/exxon-silences-staff-alarmed-by-the-climate-crisis-says-former-employee-enrique-rosero.

Horrified onlookers: Geoff Dembicki, "Exxon's Massive Offshore Oil Project Is a 'Carbon Bomb': Environmental Group," *VICE News*, September 1, 2020, www.vice.com/en/article/qj4a97/exxon-oil-guyana-climate-change.

A later investigation by Inside Climate News: Nicholas Kusnetz, "Exxon Touts Carbon Capture as a Climate Fix, but Uses It to Maximize Profit and Keep Oil Flowing," *Inside Climate News*, September 27, 2020, https://insideclimatenews.org/news/27092020/exxon-carbon-capture/.

and he saw one document as particularly disturbing: "1998 American Petroleum

Institute Global Climate Science Communications Team Action Plan," www.climatefiles.com/trade-group/american-petroleum-institute/1998-global-climate-science-communications-team-action-plan/.

a 2017 peer-reviewed paper about Exxon from Harvard researchers: Geoffrey Supran and Naomi Oreskes, "Assessing ExxonMobil's Climate Change Communications (1977–2014)," *Environmental Research Letters* 12, no. 8 (August 23, 2017), https://iopscience.iop.org/article/10.1088/1748-9326/aa815f.

Exxon reacted by accusing Oreskes and Supran themselves: "Understanding the #ExxonKnew controversy," Exxon, February 10, 2021, https://corporate.exxonmobil.com/Sustainability/Environmental-protection/Climate-change/Understanding-the-ExxonKnew-controversy.

Rosero began posting about climate change on internal Exxon message boards: Dembicki, "How Exxon Silences Staff Alarmed by the Climate Crisis."

"Is there risk?"

Trudeau unveiled an aggressive-sounding plan: "Trudeau Plan to Increase Canada's Carbon Tax Draws Mixed Response," *Al Jazeera*, December 11, 2020, www.aljazeera.com/news/2020/12/11/trudeau-pledge-to-increase-carbon-tax-draws-mixed-response.

But the reality was less than inspiring: Renaud Gignac and Dave Sawyer, "Canada's Carbon Pricing Update Improves Certainty, but Neglects Industrial Emissions," Canadian Institute for Climate Choices, December 9, 2021, https://climatechoices.ca/canadas-carbon-pricing-update/.

Trudeau made a last-minute attempt to convince Biden: Reuters Staff, "Canada Is Pressing Biden Administration on Keystone XL Pipeline, Trudeau Says," Reuters, January 19, 2021, www.reuters.com/article/us-usa-biden-keystone/canada-is-pressing-biden-administration-on-keystone-xl-pipeline-trudeau-says-idUSKBN29O267.

Biden promised to spend trillions of dollars: Joe Biden, "Remarks by President Biden Before Signing Executive Actions on Tackling Climate Change, Creating Jobs, and Restoring Scientific Integrity," White House, January 27, 2021, www.whitehouse.gov/briefing-room/speeches-remarks/2021/01/27/remarks-by-president-biden-before-signing-executive-actions-on-tackling-climate-change-creating-jobs-and-restoring-scientific-integrity/.

Exxon announced it was eager to help: Ben Lefebvre, "ExxonMobil's Climate Pitch to Biden: A $100B Carbon Project That Greens Hate," *Politico*, April 19, 2021, www.politico.com/news/2021/04/19/exxonmobils-carbon-project-biden-483253.

Greenpeace activists posing as corporate headhunters: Chris McGreal, "ExxonMobil Lobbyists Filmed Saying Oil Giant's Support for Carbon Tax a PR Ploy," *Guardian*, June 30, 2021, www.theguardian.com/us-news/2021/jun/30/exxonmobil-lobbyists-oil-giant-carbon-tax-pr-ploy.

He described McCoy's comments as "disturbing and inaccurate": Darren W. Woods, "Our Position on Climate Policy and Carbon Pricing," Exxon, July 2, 2021, https://corporate.exxonmobil.com/News/Newsroom/News-releases/Statements/Our-position-on-climate-policy-and-carbon-pricing.

Exxon was spending hundreds of thousands of dollars on Facebook ads: Brian Schwartz and Jacob Pramuk, "Exxon Mobil Has Been Lobbying Against Parts of Democrats' Big Social and Climate Spending Bill," CNBC, September 29, 2021, www.cnbc.com/2021/09/29/exxon-mobil-has-been-lobbying-against-parts-of-biden-reconciliation-bill.html.

Charles Koch also wanted to present himself as an ally: Juliana Kaplan, "Charles Koch Doubles Down on Saying He 'Screwed Up' With Partisanship, but He's Still Supporting a Republican in the Georgia Runoffs," *Business Insider*, November 24, 2020, www.businessinsider.com/charles-koch-doubles-down-saying-he-screwed-up-with-partisanship-2020-11.

the Koch network began trying to sink the bill: Andy Kroll and Geoff Dembicki, "The Koch Empire Goes All Out to Sink Joe

Biden's Agenda—and His Presidency, Too," *Rolling Stone*, September 30, 2021, www.rollingstone.com/politics/politics-news/climate-koch-brothers-lobbying-biden-build-back-better-1234815/.

had lobbied heavily in favor of Enbridge's Line 3 project: "What do Minnesota's Refineries Say About the Need for Oil From Line 3?" Minnesotans for Line 3, last updated 2020, www.minnesotansforline3.com/refineries/.

premier Jason Kenney was thrilled the project was going forward: Gordon Jaremko, "Enbridge's Line 3 Replacement Set to Flow Oil Oct. 1," *Natural Gas Intelligence*, September 30, 2021, www.naturalgasintel.com/enbridges-line-3-replacement-set-to-flow-oil-oct-1/.

concluded Lee Fang in the Intercept: Lee Fang, "Congress Outmatched by Oil Executives at What Was Meant to Be a Defining Hearing," *Intercept*, October 29, 2021, https://theintercept.com/2021/10/29/big-oil-tobacco-oversight-hearing/.

was cut due to Manchin's opposition: Coral Davenport, "Key to Biden's Climate Agenda Likely to Be Cut Because of Manchin Opposition," *New York Times*, October 15, 2021, www.nytimes.com/2021/10/15/climate/biden-clean-energy-manchin.html.

Oswald would soon be at the center of a minor scandal of his own: Emily Atkin and Jesse Coleman, "Exxon Is Still Denying Climate Science—Including Its Own," *Heated*, December 9, 2021, https://heated.world/p/exxon-is-still-denying-climate-scienceincluding.

EPILOGUE

I got in touch with Joanna Sustento over Zoom: Joanna Sustento, interview with author, November 15, 2021.

more than 2.6 million acres of California: Gabrielle Canon, "What the Numbers Tells Us About a Catastrophic Year of Wildfires," *Guardian*, December 25, 2021, www.theguardian.com/us-news/2021/dec/25/what-the-numbers-tells-us-about-a-catastrophic-year-of-wildfires.

torrential rains that caused $450 million worth of damage: Brieanna Charlebois, "B.C. Floods Cause at Least $450 Million in Damage, Insurance Bureau of Canada Reports," *Toronto Star*, December 9, 2021, www.thestar.com/politics/2021/12/09/first-nations-call-for-better-alert-system-more-funding-after-slow-flood-response.html.

he wasn't hopeful that we'd be able to stop climate change completely: Bill McKibben, interview with author, December 13, 2021.

Enrique Rosero agreed that his former employer: Enrique Rosero, interview with author, February 11, 2022.

Index

I

J

K

L

O

T

U

V

W

Y

Z

DAVID SUZUKI INSTITUTE

THE DAVID SUZUKI INSTITUTE is a nonprofit organization founded in 2010 to stimulate debate and action on environmental issues. The Institute and the David Suzuki Foundation both work to advance awareness of environmental issues important to all Canadians.

We invite you to support the activities of the Institute. For more information please contact us at:

David Suzuki Institute
219 – 2211 West 4th Avenue
Vancouver, BC, Canada V6K 4S2
info@davidsuzukiinstitute.org
604-742-2899
davidsuzukiinstitute.org

Checks can be made payable to The David Suzuki Institute.